POLITICS AND THE ENVIRONMENT IN EASTERN EUROPE

Politics and the Environment in Eastern Europe

Edited by Eszter Krasznai Kovács

OpenBook Publishers

https://www.openbookpublishers.com

ISBN Paperback: 9781800641327
ISBN Hardback: 9781800641334
ISBN Digital (PDF): 9781800641341
ISBN Digital ebook (epub): 9781800641358
ISBN Digital ebook (mobi): 9781800641365
ISBN XML: 9781800641372
DOI: 10.11647/OBP.0244

Cover image: 'People before coal' action (Warsaw, 18 November 2013). People from around the world gathered in front of Poland's Ministry of Economy in protest of the World Coal Association's International Coal and Climate Summit organised on the sidelines of the 19th UN climate change conference. Flickr, https://bit.ly/3wumjlP
Cover design by Anna Gatti

Contents

Acknowledgements

This book became possible because of the generous support of the Leverhulme Trust through an Early Career Research Fellowship, which during my time at the Department of Geography at the University of Cambridge was also financed by the Newton Trust. The time and independence of thought won for research through such grants is incalculable. Workshops and conferences were also enabled through a British Academy Small Grant.

I would like to thank Eszter Kelemen, György Pataki, Jessica Hope, Tatiana Thieme and Bill Adams for their comments and conversations on earlier drafts and iterations of some of the arguments presented here.

I will also never forget how the final versions of these chapters took shape, as bubbles of time unpredictably won during little Sophie's naps. I am very thankful for the understanding, patience and kindness shown by all the contributors and supporters to this volume during this time.

Contributors

Renata Blumberg is an Assistant Professor in the Department of Nutrition and Food Studies at Montclair State University, in Montclair, NJ. She has a BA in Anthropology from Columbia College, Columbia University, an MS in International Agricultural Development from the University of California, Davis and a Ph.D. in Geography from the University of Minnesota. Most of her research has been on alternative food networks in eastern Europe, but she has recently started a project in the USA on efforts to improve access to farmers' markets. Her other research interests include feminist pedagogy and critical dietetics.

June Brawner is a Policy Adviser at the Royal Society of London, UK. She received her Ph.D. in Anthropology at the University of Georgia in 2019. She has conducted research in Hungary since 2010, receiving a Mellon-Council for European Studies Dissertation Completion Fellowship (2018–2019) and Fulbright Study Award (2016–2017). She earned an MS in Crop and Soil Sciences from UGA (2018) and an MA in Sociology and Social Anthropology at the Central European University in Budapest, Hungary (2011). She has worked and published on many interdisciplinary projects related to food justice, human rights, rural development and musicology.

Mikuláš Černík is a Ph.D. candidate at the Department of Environmental Studies of Masaryk University in Brno, Czech Republic. His research focuses on resistance to coal mining in the Czech Republic and Poland. He is also involved in the climate justice movement *Limity jsme my*, and is a member of *Re-set,* a platform for socio-ecological transformation.

Alexandra Coțofană's research explores intersections of politics, modernities and ontologies of governing. Alexandra's scholarly interests focus on political ecologies, the ontological turn, the study of political elites and ways of governing, as well as the occult as a tool for governing, and discursive techniques employed in populist imaginaries to form racial, gender, and political Others.

Jovana Dikovic is a social anthropologist currently running her postdoctoral research project at the Department of Social Anthropology and Cultural Studies of the University of Zurich, where she also teaches. In her career, she has focused on and studied the rural Balkans. Her research and publications are mainly concerned with understanding change in rural areas, the ethics of production and soil, institution building and cooperation. In her ongoing project 'Farming under barricades: Study of cooperation in post-conflict Kosovo', she analyses farming in the context of post-war institutional anguish and stabilisation of inter-ethnic relationships.

Jana Hrckova is a Ph.D. Candidate at the Department of Sociology and Social Anthropology at the Central European University in Budapest. Her current research focuses on air pollution and green infrastructures in Warsaw.

George Iordăchescu currently researches illegal logging and the timber trade as part of the BIOSEC Project at the Department of Politics at the University of Sheffield, UK. He approaches forest crime from a political ecology perspective while tracing the socio-economic entanglements of deforestation and related illicit businesses in Romania and the EU. Over the last few years, George has undertaken fieldwork in the Romanian Carpathians and Poland, focused on private conservation projects, customary land governance and forest livelihoods. His Ph.D. thesis, completed at the IMT School for Advanced Studies in Lucca, investigated the production and protection of wilderness in eastern Europe. In the coming years, George will start a new research project investigating the illegal wildlife trade in European species, in which he will combine approaches from political ecology and green criminology.

Eszter Krasznai Kovács is a Lecturer in Environment, Society and Politics at University College London. Her research looks at how conservation and environmental management are thought about and realised between 'centres' and 'peripheries', such as between the urban and rural, the European west and east, the Global North and South. She has a particular interest in farming and food systems and agricultural policy's effects on biodiversity, water and farmers' livelihoods.

Balsa Lubarda is an environmental sociologist and a doctoral student at the Department of Environmental Sciences and Policy, Central European University (Hungary). His research explores the ways in which the far right engages with environmental and agricultural politics in post-socialist realms. Balsa is principally interested in how ideological amalgamations, conveyed through metaphors and framing, mobilise social action.

Éva Mihalovics is a Ph.D. researcher at the Geography Department of Durham University. Her Ph.D. project, 'A Quest for the Local in the Hungarian Countryside', aims to explore how different ontologies and different understandings of other-than-humans become recognised and acknowledged or ignored and denied in sustainable agricultural social enterprises in rural Hungary. Éva deploys a decolonial approach to the CEE (semi)peripheries in her work.

Arnošt Novák is an Assistant Professor of Environmental Sociology at the Faculty of Humanities of Charles University in Prague. His main research interests include environmental movements and environmental politics, commons, urban activism, and autonomous politics. He is the author of *The Dark Green World: Radical Ecological Activities in Czech Republic After 1989* (SLON, 2017, in Czech). His texts have appeared in *Journal of Urban Affairs, Social Movements Studies, Communist and Post-Communist Studies* and *Baltic Worlds*.

György Pataki is a Senior Researcher for the Environmental Social Science Research Group (ESSRG), an independent research SME. He belongs to the European Society for Ecological Economics (ESEE) and is interested in diverse research streams ranging from biodiversity and ecosystem services to solidarity economy and social innovation. He is committed to participatory and action-oriented research methodologies and practices.

Imola Püsök, University of Pécs, Hungary, is currently a Ph.D. student in anthropology, carrying out research in villages that are a part of the commune of Roșia Montană, with a special focus on Corna. She is interested in (post)industrial communities and post-socialism, modes of living, economic anthropology, dwelling and the landscape, the production of space and place and understanding change.

Introduction
Political Ecology in Eastern Europe

Eszter Krasznai Kovács

Eastern Europe is often seen as a 'lab' or model for economic and socio-political experiments of "late industrial modernity", with the region's political agency muted by capitalist and socialist planners alike (Bockman and Eyal, 2002; Glassheim, 2006: 89). This volume takes as its starting position that there are valuable things to learn *from* the eastern European (EE) region despite (or because of) these silences and repressions (Rogers, 2010). Our edited collection explores the contemporary dynamics of environmentalism across EE between people, the environment and the state, identifying associated movements, practices and relations, and sketching out contemporary systems of politics and affect. We take as fundamental a 'political ecology' approach, where nature and ecology are understood to be produced through politics, culture and history, and the political is understood as both made through the material environment and as having environmental consequences (Gille, 2009). In doing so, we contribute to emergent understandings around the pivotal role of environmental resources in the development and maintenance of new political configurations, while also providing insight into how society-nature relations are a part of the region's conceptions of place, landscape, livelihoods, and present and potential political futures.

Typical depictions of socialist legacies consist of largely grey urbanscapes, dirty streets and buildings, people queuing for basic goods, enormous coal smokestacks polluting the sky, vast expanses of decimated brown-mining sites, intensively managed agricultural areas, and, of course, the spectre of Chernobyl (Pavlínek and Pickles, 2000).

 https://doi.org/10.11647/OBP.0244.14

The 'real' environmental record of the state-socialist system is far more nuanced: alongside its share of environmental disasters and inefficiencies, the mosaic nature of intensive resource extraction also led to the persistence of a number of unique habitats; the highest forest coverage and levels of biodiversity found on the European continent; and a number of domestic (such as vegetable gardening, livestock-rearing) and societal norms that privileged self-sufficiency and local economies in ways enviable to contemporary environmental 'small-is-beautiful' movements (Smith and Jehlička, 2013). For example, the socialist system maintained and encouraged a distinct frugality borne not only out of a fundamental scarcity of goods, but also a waste processing system that mandated repair, re-use and waste recycling (Gille, 2007). These practices serve as a strong contrast to Western capitalist norms and question the 'wasteful' and 'inefficient' depictions of socialist eastern Europe (Gille, 2010; 2007).

These examples speak to the intertwined nature of the environment and society, as well as the existence of an orientalised 'ecological East' that was "invented" during the Enlightenment and further concretised by the region's post-WWII Soviet occupation and socialist decades (Wolff, 1994; Snajdr, 2008). The chapters that follow interrogate these themes through close ethnographies and engagements with local, extant forms of environmentalisms in the EE region today. The chapters' value lies in their sensitivity to the contradictions of environmentalism and the 'messy' means by which these can be realised, the increasing ways environmental issues overlap and intersect with other 'tricky' issues such as social justice, and the 'wicked' dilemmas conservation and environmental management pose on the ground.

The contradictions of environmental movements are many and shared globally, as modern-day political economies give rise not only to profligacy and waste but extensive socio-economic inequalities. The highly political questions raised by environmentalism are in the spotlight here, from how environmentalism is realised, to what it represents, who articulates it, and how the state reacts. This politics of engagement is relatively new in the EE context, as environmental groups became formally institutionalised only from the early 1990s (Herrschel, 2007; Smith and Timár, 2010; Smith and Jehlička, 2007). From this time, environmental groups were rapidly politicised as they engaged with wider political questions, particularly concerning public decision-making (or its outright lack), or became used for the expression of a nativist nationalism (Auer, 1998; Dawson, 1996; Gille, 2009; Hicks, 2004).

The political engagement of the environmental movement is often a point of critique, given its self-depiction as founded on science and objective environmental 'problems'. Although environmental protests are frequently grounded in discrete and identifiable scientific issues or localised causes, the remedies or trajectories for improving environmental conditions are ultimately reliant on political decision-making pathways that, through their variety, are by definition subjective. This subjectivity means that environmentalism often splinters as those affected or participating coalesce around different solutions. This means that environmentalism may lose its representativeness over time, or potentially even its closeness to its original 'cause' (e.g. Snajdr, 2008 describes these processes in Slovakia; Smolar, 1996). It is also arguably inevitable that a movement that often (especially in its 'deep green' incarnations) agitates for systemic change—to curb consumption and economic growth, or limit climate change—should become embedded *amongst* a host of issues rather than remaining as a separate entity. Environmental issues underpin the workings of and blur the boundaries *between* different sectors (Dalton, 2009). By its nature, therefore, environmentalism is realised through a fluctuating set of stakeholders, whose perceived acceptability, legitimacy, and thus power are constructed by wider political machinations. Today, we are at a juncture where 'ecology' as a field of knowledge—and the wider scientific and research process—is facing serious communication and credibility challenges in the midst of increasingly politically polarised, populist systems (Neimark et al., 2019). In this context, the eastern European region can perhaps lay claim to the dubious merit of setting precedents.

This volume thus grapples with the broad political 'place' of environmental issues, their reception and its spokespeople at a number of scales. The 'fall' of socialism post-1991 and eastern European accession to the EU post-2004 meant not only the passive downloading of environmental legislation and the new headquartering of Western conservation agencies in the region, but also the import of a number of development expectations, legal norms and processes. One example has been the expectation—held by Western financing and political institutions—of development of a so-called 'civil society' or 'third sector', which were formally absent within the state-socialist system (Torsello, 2012). A vibrant, active civil society is viewed as essential to democracy in the West, as part of a desirable social order that often pits civil society against the state (Urry, 1981). Assumptions around the possibilities of 'mediating' institutions are problematic, as such ideal or universalist categories fail to capture societal relations and power realities in non-Western societies (Gledhill, 2000; Hann, 1996). For example, the acceptability of CSOs operating in opposition to the state is today of enormous contention in the region, particularly in Poland and Hungary, while there is low societal trust towards these institutions and high foreign aid dependency in Bulgaria and Romania (Vandor et al., 2017). The 'foreign' nature of civil society institutions, as well as the environmental ideas and ideals that they purport to represent, can form the prongs of populist attack against them.

Providing insight into how these somewhat grandiose ideological battles play out on the ground is the undertaking of anthropologists and social scientists, who explicitly work to decentre Western constructs and assumptions around norms (Verdery, 2003). Indeed, as Doug Rogers outlines, the "very act of trying to create them [these norms and institutions] permits a critique of their assumed universality, transportability, coherence, or desirability" (Rogers, 2010: 7). For these reasons close attention is paid to the informal and interpersonal practices, the cross-institutional linkages between 'sectors' in this volume (Hann, 1996). Scholarship from the region has only recently come to focus on the role of 'nature'—how it is enacted, treated or protections and interventions configured and realised—in emergent, and now increasingly consolidating, political architectures in EE.

In the following section, an elaboration of regional political ecology is explored, including a critical look at the regional grouping and meaning of 'Eastern Europe'. While markers of comparison "along externally derived and allegedly universal metrics of, for instance, freedom, corruption, marketisation, or the development of civil society" (Rogers, 2010: 12) are imbued with Western norms, expectations and external dependencies, there is enormous merit to exploring the present political realities of the region through reference to EE's unique histories, local environment-society and power relations, and their global connections through inquiry that is sensitive and privy to local 'ways of doing things' (Burawoy et al., 2000).

A Regional Political Ecology

Political ecology focuses on the relations between 'nature' and people through attention to how natural resource access is governed from local to global scales, where environmental 'problems' are recognised as "social in origin and definition" (Watts and Peet, 2004: 7). The discipline has a strong focus on analysing the workings of power (in terms of the forms it takes, who articulates and wields it), and the differential consequences of intervention and management regimes on stakeholders *and* the environment (Robbins, 2008).

Bringing together the political and the ecological means recognising the agency of both people and the environment. Environments influence social realities: of local identities, livelihood possibilities and resources' own management. For example, Staddon (2009) explores how particular trees and forest stands are "complicit" in some forms of tree theft in Bulgaria, whereby trees' environments lead them to not 'behave' and grow as would be convenient to their "masters" — and scientific forestry. Recognising non-human agency also means re-focusing scholarly attention to perspectives and things normally unseen and without representation. These recognitions are important as the ways in which the environment is discussed matters: interventions are justified and legitimised through narratives and discourses that usually emanate from those who are in power, who decide management pathways (Forsyth, 2004).

These roles and agencies between people and the environment also transform citizen-state relationships. Centralised control, ownership and planning architectures were rapidly eroded in EE in the post-1991 period, leading to new rules and networks of resource extractors, owners, users, and so on. Newly-established EE democratic states privatised previously nationalised companies and industries in the aftermath of the collapse of intensive extraction of resources (coal, bauxite) and farming systems (arable and fruit, through cooperatives) (Oldfield, 2017). Conservation and environmental interests emerged and spread during this same period. These interests "re-territorialised" (Adams et al., 2014) claims to land, its uses and value, and have proven conflict-ridden (Petrova, 2016). Land users' ways of working and motivations have also been transformed: from 2004, the EU's CAP has incentivised farmers into a new 'service' class (Heatherington, 2011; Kovács, 2015), as governments have also broadly aimed to create new middle classes (Buchowski, 2008; Hann, 2019). Such transformations have been encouraged through economic policy and sustainable development discourses that are 'neoliberal', which expressly prioritise the development of entrepreneurial individuals and entrench market-based solutions to public policy issues. These approaches do not contest problematic power hierarchies, nor high extraction and environmental degradation rates (Staddon and Cellarius, 2002).

Political ecology seeks to understand the differential effects of environmental, social and political-economic change on different societal groups through attention to their particular life experiences and perspectives. The typical lens through which political ecology inquiries present their analyses is 'neoliberal', in that it considers how 'successfully' or 'unsuccessfully' particular groups have navigated new systems. It subsequently brands these individuals or groups to be either 'winners' or 'losers'. This prism is "in virtually all elite discourses about the masses", and is both "simplistic and ethnocentric" (Sampson, 2009: 220–21). It is important to keep in mind that "ostensible 'winners' are constantly afraid of 'losing' and that those seemingly excluded from the economic development of EU integration still pursue strategies whereby they, too, can 'win'" (ibid.), meaning that the contradictions and nuances of everyday lives need to be foregrounded.

For these reasons, many of the chapters in this book approach this challenge through locally engaged ethnographic research, to understand how people make meaning from and through changing politico-environmental contexts, circumstances and opportunities. They focus on a wide range of participants in this panoply: from particular environment action groups and political institutions to state responses, individuals' and sectors' identity shifts and transformations.

Over the past three decades, environmental resources have become a key battleground and point of contestation between local and international stakeholders, from communities, to corporations, to states, as each seeks to secure and formulate their needs, access and dominium. Recent electoral results within several EE states have had enormous consequences for the forms of control and legal regulations over these resources, particularly as we experience an apparent decline in European 'soft power', and increasingly authoritarian domestic practices. The emergence of a new authoritarianism has given rise to a re-centralisation of decision-making pathways, and erosions in the operating space of civil society groups. More broadly, several governments have instituted high levels of corporate capture and corporatised government institutions through the reassertion of politically centralised monopolies, as in Hungary and Poland. Other states have established economic competition and brokerage systems for public policy issues, as for example in Czechia, Slovakia and Romania (Innes, 2014; see also; Stahl, 2012). At the same time, these political developments often encourage and normalise a resource nationalism that justifies regulatory interventions for the control, nepotistic privatisation or re-nationalisation of natural resources, and the protection and promotion of particular landscapes (Koch and Perreault, 2018). These developments have stark, long-term consequences for society and the environment and force a strong reckoning with where eastern European societies—and thus the European Union as a whole—are headed in terms of their political democracies, economic systems and environmental futures.

Environmental struggles are in and of themselves not unique to the post-socialist region (Petrova, 2016). It is thus important to contextualise the continued relevance of the post-socialist category: after all, the collapse of socialism was over thirty years ago, meaning that many adults in the region today do not have lived memories of state-socialist regimes. The region is also well past the moment of new beginnings; indeed, barely twenty years had passed when surveyed eastern Europeans expressed widespread discontent with the 'new' world order and their place in it, and 1991 has come to be viewed by many as a "failed point of origin" and lost opportunity (Nadkarni, 2020: 3).

Political ecology in its post-structural turn questions the seemingly 'natural' delineations or representations of territory and space (Escobar, 1996). Intra-regional (internal) insistence on an eastern European 'difference' is perhaps most often grounded within claims to a (still desired) cultural difference that originates from alternative experiences and interpretations of historical events and their consequences. The historical case for the delineation of 'Eastern Europe' as separate is made with reference to the longevity of orthodox Christianity and the form of feudal relations in the region that saw power dispersed between 'noble' land-holders, in contrast to the more centralised power relations that existed between nobles and the monarch in the West (Rady, 2000). This difference, it is argued, led to the West's earlier development and institutionalisation of an 'absolutist' state, which consisted of the normalisation of legal norms and population-state relations. The lack of these relations in EE justified the label of "backwardness" to the region (see for instance Chirot, 1991). This 'backwardness' set the stage for eastern Europe to adopt an "imitative strategy of catch-up" relative to the West (Berend, 1986: 329), in an effort to emulate, join and be recognised as equal European partners on political, economic and social terms set by the West.

This suggests that "orientalising" discourses and actions arise *from within* eastern Europe as well as from without (Zarycki, 2014). Orientalising discourses and ideologies result in east/west binary thinking that has economic and power consequences, most notably that the East is on a path-dependent course to become like the West. This view holds Western capitalist systems as exemplars, and problematically relegates everyone and everything else to irrelevance, eliminating non-Western agency. EE states and actors have a significant hand in how these power relations are made manifest—from acting as imitative "servile boot-lickers" to the 'First World' and "paternalistic companions" to the 'Third' (Nowak, 2016, in Grzechnik, 2019).

Bringing EE countries "in from the cold" through formal memberships of international alliances such as NATO and the EU has not necessarily thawed differences or closed gaps. Some scholars suggest that the "EU expanded its empire" rather than undergoing proper 'integration', where integration "implies mutuality, exchange, equality," (Böröcz, 2001; Grzechnik, 2019: 1002; see also Zielonka, 2006). Instead, EE states remain crucially outside key decision-making circles. Merje Kuus' work within the European Commission found an "orientalist discourse that assumes essential difference" between eastern and western Europe, where the east is "at a distance from" the West and lacks "Europeanness", as manifest in their different styles and modes of diplomacy (Kuus, 2004: 472). In practice, this has meant that regional leaders have had little power to influence the terms of their participation in policies that they have had to implement (Caddy, 1997; Kuus, 2007; Böröcz interview to Heimer in *Népszava*, 2019). Domestic 'downloads' of EU policies have been cached in technical language, with the political (and economic) consequences under-examined or entirely smoothed over (see Dimitrova, 2020 for an exploration of these issues in Bulgaria), which allowed integration to proceed despite the threat of deadlock. The process of negotiations and preparation, however, could be seen as a constant switching between the technical parts of the acquis and their potential. When considered together, the persistent side-lining of the EE region from political processes, regional EE sovereignty concerns regarding EU-power creep, and the decontextualised branding of EE economies and democracies as perpetually behind may in part explain the current reactionary populist, nationalist turn taking place within the EE region, which risks undermining the EU project itself (Mälksoo, 2019).

These processes—the expected passive acceptance of European laws and regulatory systems with no eastern input; the righteous need to 'transform', 'develop', 'democratise', or 'lift up' the East—demonstrate how the EE region continues to be viewed as an 'underdog', forever lagging behind politically and culturally, in need of socio-economic acceleration, on terms and conditions set by the European Commission and the EU's triad of financial institutions.

There is thus a case to be made for the relevance of the EE or indeed the post-socialist category because it is lived and enacted as difference, from within and without, while at the same time being alive to the geographies and power relations of externally determined (yet internalised) markers of progress (Tlostanova, 2015). The themes of emulation, or outright dependency, run through this volume within the environmental sphere. However, the chapters do not dwell on dependency frameworks: they consider instead local assertions of ideology, activism and policy, to reinscribe agency to local actors, civil society and political groups and their struggles. Attention to local experiences contests the dominant normative expectations concerning the region's economic and civic development: paying heed to socialist continuities and post-socialist relations provides alternative interpretations and insights into why things are the way they are, and how they may proceed from here (Stenning and Hörschelmann, 2008; Verdery, 2003).

This book offers a different vantage of EE through the use of political ecology, to show how on-the-ground experiences and histories of place contest dominant narratives and experiences of unequal politics, and how, in moments of encounter and contention, alternative 'natural' environments may be brought into being. High legislative activity around natural resources directly affects a large proportion of EE populations, as post-socialist societies' demographics are still majority based in small towns or rural areas, such that many livelihoods are still (or once more) linked to resource extraction, industry and manufacturing sectors. In these contexts, environmentalism poses both a threat to traditional livelihoods and political allegiances, and opportunities for meaningful community engagement, local expressions of value, and the development and maintenance of already sustainable practices.

This Book's Sections

This collection was born from an informal early-career research network whose members met through the years at a number of political ecology conferences, where we held panels on the regional emergence of populism and populist governments. We have each, in our various ways, been engaged in an effort to better understand and document the ecological and political consequences of these turns. Research that investigates the populist tropes, minutiae of democratic decline and normalisation of authoritarian practices is crucial, but threatens to obscure the impact of 'everyday politics' (Kerkvliet, 2009) on the lives of individual citizens, the real audiences for whom national governments 'perform' their rhetoric and diplomacy. The chapters in this volume try to strike a balance between broad environmental movements, civil society structures and state responses (the first half of the book), and local-level transformations, where 'high' politics are referenced only fleetingly, or glancingly, yet influence circumstances and contexts as they develop. This twofold approach aims to provide a better understanding of the ways in which new systems of governance and power are normalised and navigated.

Questions of environmental resource access and control are fundamentally intermingled with the politics of the day in EE at a number of levels: state centralisation of power increasingly determines the ways in which any civil society may exist and operate; the oligarchisation of society and the emergence of new middle classes brings resources into new ownership patterns with untold consequences for local users or the state of the resources themselves; environmental actions risk sidelining other ways of valuing, using and seeing the environments requiring protection.

This volume is divided into three parts. **Part I** engages with the development of formal environmental movements in EE through civic actions or civil society organisations and their tactics, together with more recent government attitudes to these developments in Czechia, Hungary and Poland. **Part II** shifts focus to start to examine how experiences of landscape and daily life give rise to an environmental politics (formal and informal) that is increasingly nationalistic in Poland and Romania. **Part III** delineates more concrete interventions or changing legal frameworks to achieve conservation, or rural development, in order to question and investigate the impact of these interventions on communities, identities, and environment-dependent livelihoods.

In **Part I**, Eszter Kovács and György Pataki document the deliberate dismantling and subsequent decline of the environmental sector in Hungary, which was once composed of research institutions, designated government departments, and civil society groups and actions. While the disarray and absence of these institutions today is largely a result of the Orbán regime, it is worth noting that previous governments and accession prescriptions from the European Union have also played a part in this dismantling of the sector. Kovács and Pataki's chapter pinpoints the ways in which the state positions 'green' interests as illegitimate and oppositional to national interests in public discourse, providing a contemporary insight into the working of the Orbán regime and how it wields its power. This contribution therefore underscores the difficulties of 'developing' or seeding environmental causes or a third sector through external financing or requirements, and highlights the need for local agency, buy-in and involvement in environmental (and other social) issues for the long term.

In contrast, in Chapter 2 Arnošt Novák unpacks the long-arc dynamics of protest and environmentalism in the Czech Republic. He documents new politicisations, radicalisations and allegiances through increased international cooperation between environmental and climate change activists that contests the historical fragmentation within the domestic post-socialist environmental movement. In contrast to the Hungarian case, Novák's embeddedness in the Czech movement allows him to pinpoint how recent largescale public campaigning has re-energised the environmental sector across Czech society, leading the formal state to reconsider its coal and climate policies. Here, civic action and the civil society sector, through their strong engagement and representation amongst the people, have power. Mikuláš Černik in Chapter 3 considers the evolution of environmental resistance developed by Novák through a very contemporary form of activism, the organisation and tactics of 'climate camps' between Poland and Czechia. Mass mobilisation that redirects states' energy policies, and deliberative state responses to these demands, provide nascent insight into how environmentalism may gain broad societal legitimacy and its own political power.

In Chapter 4 Jana Hrckova also documents a contemporary example of activism or, using her term, 'activation' by local Warsovian residents to 'save' green spaces in Warsaw. Residents' actions have challenged the neoliberal urban planning rhetoric of town planners and city administrators intent on turning Warsaw into a 'world-class' city worthy of Western investment. Helped along by a number of political scandals and crises at the city level, the result of local agitation may not only be the preservation of green sites across the city, but the creation of new imaginaries and "manoeuvring space" for what a city should look and feel like, and how it can be lived in and used. Hrckova develops the notion of how technical language is often hollow, allowing it to be inverted and deployed by environmentalists to suit preservation agendas that are diametrically opposed to the construction-heavy development prerogatives of the typical post-socialist city trying to compete on a global urban stage.

These chapters together offer compelling, multi-scalar reflections on the space an environmental 'civil society' has been able to carve out and operate within, in political contexts that often view this sector (both civil society and environmentalism) as antagonistic to state interests. This is most dramatic (although at the same time subtle and accumulative in its methods) in the Hungarian case, but the uses of violent state institutions against environmental activists is an increasingly global phenomenon (Feng et al., 2020). The research from Czechia and Poland demonstrates the importance of international connections for local civil movements, as sources of regionally novel forms of activism (climate camps, coal protest), solidarity, and also as reminders of the borderless nature of many environmental issues. At the same time, these studies document a spectrum of state tactics to police, surveil or outright suppress environmentalism, where the branding of environmentalists as 'extremists' gives state institutions enormous (legal, as well as physical, intimidatory) power to curb and cow these movements—but where such labels might also legitimise and bring together formerly fragmented groups. All of these chapters demonstrate in different ways the merit of looking at the connections and networks behind and within environmental movements, as the various actions taken by groups in this sector may find potentially unexpected but valuable alliances in one another. The chapters also point to a fragility in the contemporary moment, a sentiment of by-the-grace-of-the-state tolerance of the existence and campaigns of environmental groups.

Part II examines the claims to alliances between 'nature, blood and soil', wherein people make meaning and politics from the contemporary landscape around them. In Chapter 5, Balsa Lubarda examines how, in Poland, ideological content is embedded and amended through activism and engagement with the fate of 'far-right' political organisations and individuals and their versions of environmentalism. Lubarda's concept of 'far-right ecologism' makes explicit the constellation of actors and environments that meld worldviews and materialities into a political position, where right-wing and left-wing environmentalists increasingly align with a return to or privileging of local protest and local issues. Nationalist claims about 'blood and soil' are explored in Chapter 6 by Alexandra Coțofană, who engages directly with the forms of nationalism expressed online and in print media by people and groups threatened by the entry of foreign interests into Romania. In her account, the Carpathian Mountains are symbolically imbued with a xenophobic agency to 'protect' those deemed to be ethnic inhabitants of the region. This agency is political and reaffirms nationalist tropes around rights to land and belonging. Also focusing on Romania and the Carpathians, in Chapter 7 Imola Püsök recounts the ways in which inhabitants of villages 'left behind' by all forms of change experienced in the region over more than three decades—including economic collapse and shifts in industrial and primary resource extraction, rural abandonment and an increasing geographical marginality, and the entry of new corporate and environmental interests—have left the region's oldest inhabitants in an increasingly precarious periphery. New modes and values of productivism brought in from 'outside' (the West, the young, the city) have displaced older conceptions of work and the factors that previously underpinned belonging. These 'foreign' agents and ideas include environmental claims by (local and international) NGOs, the extractive aspirations of gold-mining companies, and control-seeking state agencies—all of which share an intent to reap, and introduce new ways of reaping, profit from nature. They are also often explicit about the need to physically displace local people to achieve their goals.

Part II's chapters highlight under-explored reasons for and forms of local nationalistic sentiment. These nationalisms originate in strong attachments to place and often an appreciation of nature that tries to also naturalise the people who inhabit these places. In the present era of globalisation, these attachments cause friction with the agents of change that are seen to arrive through external sources, from resource extraction processes, or development and infrastructure projects, or as security forces from distant states. Local knowledge of the natural environment, together with historical claims, strengthen nationalistic convictions, and suggest that increased populist rhetoric by *state* authorities reflect as well as seed pre-existing sentiments of dispossession and othering. However, as Lubarda also notes, localism as expressed through "ecological forms of nationalism" may be "potentially progressive and desirable", as local duties and responsibilities may give rise to non-exclusive ethics of care.

There is a rich vein of scholarship on the EE region concerning the continuities and transformations in local natural resource access and use customs. This work has focused on intimate local knowledge of landscapes arising from a number of activities on the land: from foraging, especially of mushrooms (where Poles are at the 'top end' of the "mycophilia spectrum", Kotowski et al., 2019; Kovalčík, 2014; Šiftová, 2020); from vegetable gardening and the continued 'embeddedness' of local food growing, sharing and markets (Jehlička et al., 2020; Jehlička and Daněk, 2017; Smith and Jehlička, 2007); from high winter fuel dependency as a result of high rural poverty (Buzar, 2007; Staddon,

2009), and the regional preponderance of largely undocumented forest extraction practices (Vasile, 2019). There has also been a continuity in the function and access of some local-level, community institutions that are at odds with more 'macro' legal changes (Cellarius, 2004). This makes the intersection between 'top-down' regulatory regimes and local customs and access an important point of critical inquiry, particularly as the persistence of local use regimes are often negotiated around or in spite of paper-based, formal systems.

Part III develops insights into these local practices, but also makes the conflict between regulatory and interventionist expectations, and local ways of doing things, explicit. The work in this section documents the ways in which local communities and practices have been disrupted by or have deliberately accommodated pressure from external financiers and private actors. When these initiatives go on to fail, there is then a reconsideration as to what went wrong, and what has changed or been left behind. The chapters document how particular Western ideas around development or environmental interventions have travelled, and document their messy fates at the local level in EE: George Iordăchescu considers conservation in Romania, June Brawner explores the transformation of Tokaj wine-making practices in eastern Hungary, Renata Blumberg emphasises the academic blinkers that fail to 'see' and learn from the value of extant local food practices in the east, Jovana Dikovic contests what success means within a rural region, and Éva Mihalovics discusses the structural and personal inequalities that set some people and projects up for failure in rural Hungary.

The people with whom we have engaged and written about in these chapters are frequently fragmented from one another and centres of power, and find themselves up against big legal-corporate machines. They cannot be completely excluded or left behind because they, their possessions and labour are required for continued production. However, local acceptance and modification of newly introduced ways of doing things have not gone smoothly, just as identity changes and the emergence of *Homo oeconomicus* were not inevitable with the collapse of socialism (Dunn, 2009). The chapters in this section also canvass how new production methods or environmental interventions in eastern Europe rely on, import and depend on Western ideas, actors, money, and so on. In Chapter 8, George Iordăchescu recounts the transposition and contemporary evolution of fortress wilderness conservation in Romania, where wild nature is emerging as an environmental fix that generates capital. National-park approaches are the prerogative and modus operandi of conservation from its earliest days in the US, and have also been adopted by the EU. Today the prospect of touristic and other capital is a motivating factor for making inroads into the Carpathians of Romania as a 'wilderness' frontier. There is a stark scalar fragmentation of reactions and adoptions of this new conservation between formal institutions of the state, the (private) agencies tasked with realising the conservation area, and local land users. This chapter points again to the sources of disenchantment with top-down civil and state projects.

June Brawner's work in Chapter 9 presents a fascinating case study of the production of (no longer so sweet) Tokaj wine, where changing practices in wine production cater to the palate of Western consumers. This very question of taste and palate is, she argues, a modern phenomenon: the advent of 'expert' wine tasting externalised taste judgement—and found the 'traditional' Hungarian ways lacking. Modern tastes have transformed the Tokaj wine market and has resulted in the active devaluation of traditional methods and land users. Brawner also questions the extent to which 'local tradition' has been eliminated or eroded by new values and processes of wine-making. Renata Blumberg in Chapter 10 takes up several of Brawner's themes through a theoretical consideration of the ways in which research ideas travel: from the popularity, and even foundational role, of eastern European agrarian societies for the study of the political economy of food systems, to the region's contemporary peripheralisation as a source of theory. Blumberg proposes a geographical approach to a political ecology of food systems that centres and recognises the many extant values and practices in EE that act as the envy of parallel 'green' or sustainability movements in the West, such as the persistence of local markets, household-level production and food storage.

Renata Blumberg also cautions against reducing all analyses of effects and processes to capitalism, as the contextual social and cultural logics of work, labour, production, and held and translated values must also be accounted for. The final two chapters approach the socio-cultural context of rural development with this objective. In Chapter 11 Jovana

Dikovic examines what constitutes 'development' and 'success' from the local perspective of land users and those dwelling in villages across Serbia. Her villagers' accounts challenge the assumptions that belie the widespread use of economic incentives to influence land-holders throughout the EU, or show what may be at stake if CAP subsidies are introduced in the near future. These incentives supposedly reorient productivism and work to more efficient land uses. However, Dikovic asks whether such exclusively economic rationales fundamentally misunderstand what motivates land-users, who are keenly connected and sensitive to their neighbours and wider communities, and motivated by more than capital accumulation.

Éva Mihalovics in Chapter 12 questions why, from her experience as a practitioner and researcher, development projects seemingly 'fail' in north-east Hungary. Her account pinpoints the objectification and often patronising engagement of development practitioners with local communities and individuals. Éva's discussion also highlights the clash of the local and pragmatic with the externally designed and funded requirements for participation in these development schemes. While these two chapters underscore the temporally limited success and longevity of 'development' imposed in technical ways from the outside, with little appreciation of (or interest in) local values or realities, they also agitate for more. In Dikovic's case, there is a plea for meaningful local engagement prior to the introduction of any intervention or policy that may carry enormous implications for local livelihoods and values. Éva Mihalovics's chapter effectively agitates for scholarly change—for all of us to reflect on how research and interventions intended to 'improve' local lives often replicate or give rise to new problems amongst communities.

***

Taken together, these contributions provide a snapshot of the complex entanglements that make up the EE region's 'environmentalisms'. A strong grasp of these complexities is crucial for a better understanding of how a greener, more sustainable economy and way of life may be fostered. At the same time, they speak to the importance of 'having an ear to the ground', and the continued need for locally engaged scholarship.

References

Adams, W. M., Hodge, I. D., Sandbrook, L., 2014. New spaces for nature: the re-territorialisation of biodiversity conservation under neoliberalism in the UK. *Transactions of the Institute of British Geographers*, 39, 574–88. https://doi.org/10.1111/tran.12050.

Auer, M. R., 1998. Environmentalism and Estonia's independence movement. *Nationalities Papers*, 26, 659–76. https://doi.org/10.1080/00905999808408593.

Berend, I. T., 1986. The Historical Evolution of Eastern Europe as a Region. *International Organization*, 40, 329–46.

Bockman, J., Eyal, G., 2002. Eastern Europe as a Laboratory for Economic Knowledge: The Transnational Roots of Neoliberalism. *American Journal of Sociology*, 108, 310–52. https://doi.org/10.1086/344411.

Böröcz, J., 2001. Introduction: Empire and Coloniality in the "Eastern Enlargement" of the European Union, in: Böröcz, J., Kovács, M. (Eds.), *Empire's New Clothes: Unveiling EU Enlargement*. Central European Review, Telford, pp. 4–50.

Buchowski, M., 2008. The enigma of the middle class: A case study of entrepreneurs in Poland, in: Schroder, I. W., Vonderau, A. (Eds.), *Changing Economies and Changing Identities in Postsocialist Eastern Europe*, Halle Studies in the Anthropology of Eurasia. LIT Verlag Münster, Berlin, pp. 47–74.

Burawoy, M., Blum, J. A., George, S., Gille, Z., Thayer, M., 2000. *Global Ethnography: Forces, Connections, and Imaginations in a Postmodern World*. University of California Press, Berkeley, CA.

Buzar, S., 2007. *Energy Poverty in Eastern Europe: Hidden Geographies of Deprivation*. Ashgate Publishing, Ltd., Ashgate.

Caddy, J., 1997. Harmonization and asymmetry: environmental policy co-ordination between the European Union and Central Europe. *Journal of European Public Policy*, 4, 318–36. https://doi.org/10.1080/135017697800000021.

Cellarius, B. A., 2004. "Without Co-ops There Would be No Forests!": Historical Memory and the Restitution of Forests in Post-socialist Bulgaria. *Conservation & Society*, 2, 51–73.

Chirot, D. (Ed.), 1991. *The Origins of Backwardness in Eastern Europe: Economics and Politics from the Middle Ages until the Early Twentieth Century*, First edition. University of California Press, Berkeley, CA.

Dalton, R. J., 2009. Economics, environmentalism and party alignments: A note on partisan change in advanced industrial democracies. *European Journal of Political Research*, 48, 161–75. https://doi.org/10.1111/j.1475-6765.2008.00831.x.

Dawson, J. I., 1996. *Eco-nationalism: Anti-nuclear Activism and National Identity in Russia, Lithuania, and Ukraine*. Duke University Press, Durham, NC.

Dimitrova, A. L., 2020. Enlargement by Stealth? Bulgaria's EU Negotiations between the Political and the Technical. *Southeastern Europe*, 44, 130–56. https://doi.org/10.30965/18763332-04402002.

Dunn, E., 2009. Subjectivity after socialism: An invitation to theory building in anthropology, in: Schröder, I. W., Vonderau, A. (Eds.), *Changing Economies and Changing Identities in Postsocialist Eastern Europe*. Max Planck Institute for Social Anthropology, Halle, pp. 225–34.

Escobar, A., 1996. Construction nature: Elements for a post-structuralist political ecology. *Futures*, 28, 325–43. https://doi.org/10.1016/0016-3287(96)00011-0.

Feng, J., Mildenberger, M., Stokes, L. C., 2020. Inhumane environments: Global violence against environmental justice activists as a human rights violation, in: *A Research Agenda for Human Rights*. Elgar Research Agendas. https://doi.org/10.4337/9781788973083.

Forsyth, T., 2004. *Critical Political Ecology: The Politics of Environmental Science*. Routledge, London.

Gille, Z., 2010. Actor Networks, Modes of Production, and Waste Regimes: Reassembling the Macro-Social. *Environment and Planning: Economy and Space*, 42, 1049–64. https://doi.org/10.1068/a42122.

——, 2009. From Nature as Proxy to Nature as Actor. *Slavic Review*, 68, 1–9. https://doi.org/10.1017/S003767790000005X.

——, 2007. *From the Cult of Waste to the Trash Heap of History: The Politics of Waste in Socialist and Postsocialist Hungary*. Indiana University Press, Bloomington and Indianapolis, IN.

Glassheim, E., 2006. Ethnic Cleansing, Communism, and Environmental Devastation in Czechoslovakia's Borderlands, 1945–1989. *Journal of Modern History*, 78, 65–92. https://doi.org/10.1086/499795.

Gledhill, J., 2000. *Power and Its Disguises: Anthropological Perspectives on Politics*, 2nd ed. Pluto Press, London.

Grzechnik, M., 2019. The Missing Second World: On Poland and Postcolonial Studies. Interventions 21, 998–1014. https://doi.org/10.1080/1369801X.2019.1585911.

Hann, C., 2019. *Repatriating Polanyi: Market Society in the Visegrád States*. Central European University Press, Budapest.

——, 1996. Introduction: political society and civil anthropology, in: Hann, C., Dunn, E. (Eds.), *Civil Society: Challenging Western Models*. Routledge, London and New York, pp. 1–24.

Heatherington, T., 2011. Introduction: Remaking Rural Landscapes in Twenty-first Century Europe. *Anthropological Journal of European Cultures*, 20, 1–9. https://doi.org/10.3167/ajec.2011.200101.

Heimer, Gy. 2019. Az EU-ról illúziók nélkül — interjú Böröcz Józseffel, *Népszava* 24 April 2019, https://nepszava.hu/3033198_az-eu-rol-illuziok-nelkul-interju-borocz-jozseffel.

Herrschel, T., 2007. Between difference and adjustment — The re-/presentation and implementation of post-socialist (communist) transformation. *Geoforum: Post Communist Transformation*, 38, 439–44. https://doi.org/10.1016/j.geoforum.2006.11.007.

Hicks, B., 2004. Setting Agendas and Shaping Activism: EU Influence on Central and Eastern European Environmental Movements. *Environmental Politics*, 13, 216–33. https://doi.org/10.1080/09644010410001685218.

Innes, A., 2014. The Political Economy of State Capture in Central Europe. *Journal of Common Market Studies*, 52, 88–104. https://doi.org/10.1111/jcms.12079.

Jehlička, P., Daněk, P., 2017. Rendering the Actually Existing Sharing Economy Visible: Home-Grown Food and the Pleasure of Sharing. *Sociologia Ruralis*, 57, 274–96. https://doi.org/10.1111/soru.12160.

Jehlička, P., Grīviņš, M., Visser, O., Balázs, B., 2020. Thinking food like an East European: A critical reflection on the framing of food systems. *Journal of Rural Studies*, 76, 286–95. https://doi.org/10.1016/j.jrurstud.2020.04.015.

Kerkvliet, B. J. T., 2009. Everyday politics in peasant societies (and ours). *The Journal of Peasant Studies*, 36, 227–43. https://doi.org/10.1080/03066150902820487.

Koch, N., Perreault, T., 2019. Resource nationalism. *Progress in Human Geography*, 43, 4, 611–31. https://doi.org/10.1177/0309132518781497.

Kotowski, M. A., Pietras, M., Łuczaj, Ł., 2019. Extreme levels of mycophilia documented in Mazovia, a region of Poland. *Journal of Ethnobiology and Ethnomedicine*, 15, 12. https://doi.org/10.1186/s13002-019-0291-6.

Kovács, E. K., 2015. Surveillance and state-making through EU agricultural policy in Hungary. *Geoforum*, 64, 168–81. https://doi.org/10.1016/j.geoforum.2015.06.020.

Kovalčík, M., 2014. Value of forest berries and mushrooms picking in Slovakia's forests. *Beskydy*, 7, 39–46. https://doi.org/10.11118/beskyd201407010039.

Kuus, M., 2007. *Geopolitics Reframed: Security and Identity in Europe's Eastern Enlargement*. Springer, London.

——, 2004. Europe's eastern expansion and the reinscription of otherness in East-Central Europe. *Progress in Human Geography*, 28, 472–89. https://doi.org/10.1191/0309132504ph498oa.

Mälksoo, M., 2019. The normative threat of subtle subversion: the return of 'Eastern Europe' as an ontological insecurity trope. *Cambridge Review of International Affairs*, 32, 365–83. https://doi.org/10.1080/09557571.2019.1590314.

Neimark, B., Childs, J., Nightingale, A. J., Cavanagh, C. J., Sullivan, S., Benjaminsen, T. A., Batterbury, S., Koot, S., Harcourt, W., 2019. Speaking Power to "Post-Truth": Critical Political Ecology and the New Authoritarianism. *Annals of the American Association of Geographers*, 109, 613–23. https://doi.org/10.1080/24694452.2018.1547567.

Oldfield, J. D., 2017. *Russian Nature: Exploring the Environmental Consequences of Societal Change*. Routledge, London.

Pavlínek, P., Pickles, J., 2000. *Environmental Transitions: Transformation and Ecological Defense in Central and Eastern Europe*, 1st edition. Routledge, London and New York.

Petrova, S., 2016. *Communities in Transition: Protected Nature and Local People in Eastern and Central Europe*. Routledge, London.

Rady, M. (2000) *Nobility, Land and Service in Medieval Hungary*. Palgrave, Basingstoke and New York.

Robbins, P., 2008. *Political Ecology: A Critical Introduction*, 1st ed. Wiley-Blackwell, London.

Rogers, D., 2010. Postsocialisms Unbound: Connections, Critiques, Comparisons. *Slavic Review*, 69, 1–15.

Sampson, S., 2009. Winners, losers, and the neoliberal self: Agency in post-transition Europe, in: Schroder, I. W., Vonderau, A. (Eds.), *Changing Economies and Changing Identities in Postsocialist Eastern Europe*, Halle Studies in the Anthropology of Eurasia. LIT Verlag Münster, Berlin, pp. 219–24.

Šiftová, J., 2020. Foraging in Czechia: The nation's precious hobby. *Norsk Geografisk Tidsskrift — Norwegian Journal of Geography*, 74, 310–20. https://doi.org/10.1080/00291951.2020.1851757.

Smith, A., Timár, J., 2010. Uneven transformations: Space, economy and society 20 years after the collapse of state socialism. *European Urban and Regional Studies*, 17, 115–25. https://doi.org/10.1177/0969776409358245.

Smith, J., Jehlička, P., 2013. Quiet sustainability: Fertile lessons from Europe's productive gardeners. *Journal of Rural Studies*, 32, 148–57. https://doi.org/10.1016/j.jrurstud.2013.05.002.

——, 2007. Stories around food, politics and change in Poland and the Czech Republic. *Transactions of the Institute of British Geographers*, 32, 395–410. https://doi.org/10.1111/j.1475-5661.2007.00258.x.

Smolar, A., 1996. Civil Society After Communism: From Opposition to Atomization. *Journal of Democracy*, 7, 24–38. https://doi.org/10.1353/jod.1996.0018.

Snajdr, E., 2008. *Nature Protests: The End of Ecology in Slovakia*. University of Washington Press, Seattle, WA and London.

Staddon, C., 2009. The Complicity of Trees: The Socionatural Field of/for Tree Theft in Bulgaria. *Slavic Review*, 68, 70–94. https://doi.org/10.1017/S0037677900000097.

Staddon, C., Cellarius, B., 2002. Paradoxes of conservation and development in postsocialist Bulgaria: recent controversies. *European Environment*, 12, 105–16. https://doi.org/10.1002/eet.289.

Stahl, J., 2012. *Rent from the Land: A Political Ecology of Postsocialist Rural Transformation*. Anthem Press, London.

Stenning, A., Hörschelmann, K., 2008. History, Geography and Difference in the Post-socialist World: Or, Do We Still Need Post-Socialism? *Antipode*, 40, 312–35. https://doi.org/10.1111/j.1467-8330.2008.00593.x.

Tlostanova, M., 2015. Can the post-Soviet think? On coloniality of knowledge, external imperial and double colonial difference. *Intersections. East European Journal of Society and Politics*, 1, 38–58.

Torsello, D., 2012. The "revival" of civil society in Central Eastern Europe: New environmental and political movements. *Human Affairs*, 22, 178–95. https://doi.org/10.2478/s13374-012-0016-1.

Urry, J., 1981. *Anatomy of Capitalist Societies: The Economy, Civil Society and the State*. Macmillan International Higher Education, London and Basingstoke.

Vandor, P., Traxler, N., Millner, R., Meyer, M., 2017. *Civil Society in Central and Eastern Europe: Challenges and Opportunities*. ERSTE Foundation, Vienna.

Vasile, M., 2019. Fiefdom forests: Authoritarianism, labor vulnerability and the limits of resistance in the Carpathian Mountains. *Geoforum*, 106, 155–66. https://doi.org/10.1016/j.geoforum.2019.08.001.

Verdery, K., 2003. *The Vanishing Hectare: Property and Value in Postsocialist Transylvania*. Cornell University Press, Ithaca, NY.

Watts, M., Peet, R., 2004. Liberating political ecology, in: Peet, R., Watts, M. (Eds.), *Liberation Ecologies: Environment, Development, Social Movements*. Routledge, London and New York, pp. 3–47.

Wolff, L. (1994) *Inventing Eastern Europe: The Map of Civilization on the Mind of the Enlightenment*. Stanford University Press, Stanford California.

Zarycki, T., 2014. *Ideologies of Eastness in Central and Eastern Europe: 96*, 1st edition. Routledge, Abingdon and New York.

Zielonka, J., 2006. *Europe as Empire: The Nature of the Enlarged European Union, Europe as Empire*. Oxford University Press, Oxford.

PART I

1. The Dismantling of Environmentalism in Hungary

Eszter Krasznai Kovács and György Pataki[1]

Introduction

In contemporary Hungary, environmental movements and concerns are treated by government agencies and their representatives as oppositional to mainstream politics. There is a long history of such antagonistic positioning from around the world, particularly as dominant political ideologies premise economic growth and social development through technological innovation and the commodification of nature. Most environmentalisms, particularly those that combine ecology with social justice (Sachs, 1995), question and challenge the ideology of this developmentalist mentality and the sustainability of these goals for nature and society. This chapter will explore the ways in which the state has effectively dismantled and eliminated the independent as well as public environmental sector in Hungary. Its dismantling has been achieved by the Hungarian government at a number of scales and forms: through closed, personal, targeted and insidious means that aim at the destruction of individuals and what they represent; as well as through outright, aggressive displays of violence and legal overreach that have served to intimidate and silence.

In the eastern European (EE) region environmental issues' perceived legitimacy and their relation to formal government have changed drastically over the past thirty years. During authoritarian-socialist

1 We wish to thank Eszter Kovács for her careful reading and contributions to an earlier version of this manuscript. All errors and interpretations remain our own.

 https://doi.org/10.11647/OBP.0244.01

times, both official ideology and state practice were adversarial towards any bottom-up civic activities beyond seemingly apolitical cultural, sports-based, and some leisure and recreational pursuits. Not surprisingly, environmentalism was typically realised through conservation activities that remained 'neutral' or 'apolitical', such as synchronised bird-watching along the Danube river. These 'hobby' activities did not threaten or contest political relations and state power, as they did not mobilise groups or engage in commentary on governmental actions (Harper, 2006). These divisions cemented scientific and political divides, as political decision-making was not linked with environmental consequences or risk (Pavlínek and Pickles, 2000). Authoritarian-socialist regimes demonstrated no difference from free market societies regarding their developmentalist mentality, in that nature was instrumentalised for the sake of human progress.

In the 1980s, local environmental struggles in the EE region emerged, typically related to the siting of hazardous industrial plants, and pollution incidents that endangered the health of local residents (e.g. waste incinerators and processing plants, highway construction, etc.; Snajdr, 2008; Vári, 1997). These environmental conflicts were unusual at the time because of their politicised nature, even if the apparent politics of the given conflicts typically focused on local opportunities for public participation in siting decisions (see, e.g. Faragó et al., 1989). These local environmental actions were characterised by a temporary coalition between experts of diverse kinds (ecological, medical, engineering) and local residents focused on human welfare at the local scale without explicitly advocating or articulating a systemic political-ecological critique (Vári, 1997).

The first environmental movement in Hungary that created a space for explicitly political participation was the Danube movement (*"Duna Kör"*) from the late 1980s. This environmental movement assembled a broad popular protest against the Slovak-Hungarian Gabčikovo-Nagymaros dam on the Danube. The Danube movement targeted the government's legal-political restrictions around access to environmental information, free speech and free civic organising—in short, democratic political rights. Their powerful framing provided an opportunity for all kinds of opposition towards the authoritarian-socialist regime to unite and march together (Buzogány, 2015; Kerényi and Szabó, 2006). Haraszti

(1990) considered the Danube movement as the first real movement beyond environmentalism and as an archetype of democratic pluralism (see also Corry, 2014; Reynolds, 2020).

Immediately preceding and rapidly after the collapse of the Soviet system, environmentalism moved into the political mainstream, as both an interest that required representation and defence, and as a form of 'party' politics (Szabó, 2000). The adoption of multi-party democracy meant a proliferation not only of civic organising but outright activism that was frequently led by local residents. Environmentalism thus became more widespread, with local groups attempting to influence concrete development decisions and projects that gave rise to new webs of emerging stakeholders with broader political agendas. During the parliamentary election of 1990, a 'green' agenda was integrated into the programmes of many new political parties (for example in Hungary, Mikecz, 2017; Slovakia, Snajdr, 2008).

Regime change also brought a new legal infrastructure and political space for civic movements. A process of professionalisation and internationalisation unfolded, including the appearance of Hungarian branches of international green organisations (Greenpeace, WWF) and the establishment of environmental non-governmental organisations (ENGOs) with special attention to and competence in particular policy fields (e.g. energy, mobility, waste, etc.). Harper (2006) sees this as a "post-socialist political ecology" that changed from a dissident and protest-style movement towards a "globalisation-from-below" (a Seattle-type alter-globalisation). During the 2000s, activists established explicitly eco-political organisations, most notably "*Védegylet*" (Protect the Future) and "*Zöld Fiatalok*" (Green Youth), which introduced to the Hungarian environmental scene participatory democracy, a critique of capitalism and consumerism and, at the same time, a call for social transformation towards sustainability. Beyond the performative acts of protest, two other roles were institutionalised: the role of 'citizen watchdogs' (Protect the Future coordinated a collaboration of green civil organisations for a 'civic ombudsman for future generations' that issued annual reports) and of 'think tanks' (Protect the Future coordinated a wide spectrum of experts producing policy briefs for particular policy fields for a sustainability transformation) (Vay, 2004). These worked to develop and demonstrate ecopolitical alternatives to a consumerism-based market society.

Accession to the European Union (EU) in 2004 strengthened these trends and opened up new possibilities (Buzogány, 2015). The EU's Acquis required the adoption of environmental legislation that introduced and made compulsory cross-sectoral standards and increased the political visibility and financing streams of 'green' civil society groups, which by implication increased the policy influence of ENGOs in Hungary (Cent et al., 2013; Mertens, 2013). In their historical review of Hungarian nature conservation policy, Mihók et al. (2017) claims that during the 1990s "the position of nature conservation within the state" was strengthened, "which resulted in the expansion of protected areas [...], conservation instruments, measures and actions" (Mihók et al., 2017: 69). The 'heyday' of environmental representation and activism in Hungary occurred before the 2008 Global Financial Crash (GFC), when the conservation sector (i.e. national park directorates) received peak financing directly from the state (Kovács, 2017), and the renowned environmentalist László Sólyom (ex-president of the Constitutional Court of Hungary, formerly involved in *Duna Kör*) was elected President of the Republic of Hungary in 2005.

While the GFC resulted in reduced monetary support, the election of the current Orbán-led *Fidesz*/Christian Democrat party coalition government in 2010 led to a shift in attitudinal relations towards environmentalism from state institutions and emerging public policy. Orbán is today well known for his aspirational declaration in 2014 of his desire to turn Hungary into an 'illiberal democracy'. The meaning of this undertaking in relation to natural resources and the ideological 'place' of environmental values and assets has become apparent in the intervening period.

We unpack these shifts, and how they have taken place, in three parts: first, we examine the interpersonally motivated, emotionally charged and selectively targeted operation of power directed by Orbán through the reasons for the closure of a research institute. Second, we look at the structural elimination of environmental interest representation through the silent dismissal of experts from within public institutions. Third, we recount the aggressive use of violent institutions of the state against environmental civil society groups. We argue that the elimination of the environmental sector is deliberate, an outcome of direct policy and decisions within the civil service, a manipulation of media narratives, targeted sectoral attacks and denial of environmental science and expertise. These machinations reveal insights into the workings—the exclusionary practices and logic—of the Orbán-led *Fidesz*/Christian-Democrat super-majority government. The deliberate re-orientation of formal institutions that manage natural resources has resulted in the relaxation of environmental governance regimes that then aids elite accumulation and questionable development works, and limits public accountability and oversight of these same institutions.

We begin below with an overview of illiberal democracy and theoretical conceptualisations for the composition and workings of the Hungarian state. We then position environmentalism and environmental expertise in relation to authoritarian state power.

Deliberate Dismantling and Authoritarianism

In 2014, PM Viktor Orbán outlined for Hungary a politics and goal of "illiberal democracy" (see Tóth, 2014 for an English translation of PM Orbán's speech). Also termed 'electoral democracy', in practice, this idea refers to a state that holds regular elections, but frequently violates the civil liberties of its citizens (Nyyssönen and Metsälä, 2021). Fundamentally, Orbán's vision held Hungary as a nation to be constructed and organised around a set of principles that set it apart, that defined it as 'Hungarian' (see Tóth, 2014). In this, the Hungarian nation-state's primary role was not to realise democracy or liberal principles such as freedom (or even the well-being of its citizens!), but to construct this community of Hungarianness around other central principles. These principles have proven, over time, to consist of a re-affirmation and organisation of society around the (heteronormative) nuclear family, Christian religiosity, and to insist that membership to such a society is predicated on a shared white ancestry. These elements are emphasised through a reimagining of history within education and public discourse; public policy campaigns against 'Others' who take variable forms, from foreigners to refugees; and a slew of lucrative economic and tax incentives to families to procreate, and to churches to increase their reach.

Crucially for our purposes in this chapter, we turn our attention to the *ways* in which the Orbán regime has come to realise this illiberal democracy—in practice a form of soft or hybrid authoritarianism somewhere between "defective democracy and electoral authoritarianism" (Bogaards, 2009). Fundamentally, the Hungarian state's violation of rule-of-law norms and civil liberties relies on there not being a broad public coalition that demands them, or a public that can effectively develop or protest their infringement. Far from the immediate post-Soviet period assumptions around the 'inevitable' route to the development of democratic and rule-of-law norms, Hungarian society has been greatly economically (neo)liberalised over the past three decades, while societally it has remained relatively closed. Its civil society sector is made up of numerous local groups that are again increasingly apolitical and focused on activities that are deemed to have minimal political risk, such as sport and leisure (Szalai and Svensson, 2018), as the Orbán government tries to administratively and financially reign in the activities and power of these groups away from politically sensitive subjects (Cox, 2020).

In Orbán's Hungary, public politics is undertaken through party allegiance, where there has been only one party and one (coalition) government, that of *Fidesz* and the Christian-Democrats. For the best part of this decade this government has controlled Parliament through a two-thirds super-majority, empowering them to change the constitution. In many ways, therefore, the state *is* the dominant political party, and *vice versa*. Key decisions and processes are not decided within transparent forums such as Parliament, but merely announced there after having been decided within closed party meetings. A dominant trend of this form of government has been state capture and repurpose, through which rent extraction for the benefit of the elite and emergent upper-middle classes has been consolidated (Gottfried, 2019; Innes, 2014).

This incarnation of the state has also been described as "network-building, mafia-like", akin to an octopus, with tentacles of control everywhere (Magyar, 2013). Alternatively, the state may be conceived as an "illiberal polypore state... like a fungus, it feeds on the vital resources of the previous system at the same time as contributing to its decay" (Pető and Grzebalska, 2016). Octopedal and fungal allegories emphasise how the state operates through the interstices and elides exact capture, as the agents of authoritarian order are everywhere, and their connections are hidden. While decision-making is thus made invisible and centralised to within party and elite members' interests (to borrow Magyar's mafia concept, the government is an "upper-world" with a "godfather" at the top and "adopted family" below), the processes and governance of the state are at the same time made informal and widely distributed. Comfortable survival within such a nation-state for an 'average' citizen becomes entirely circumstantial, enormously dependent on who you know, what you know, and what resources (social, financial) a person can mobilise to get by (Polese, 2014; Polese et al., 2015; Stepurko et al., 2015).

The maintenance of such a system relies on the ability of the central state to control key facets of its citizen's choices, from media consumption to political alternatives. Recent geographical scholarship has highlighted the ways in which the environmental sector has been deemed a source of challenge to hybrid authoritarianism (McCarthy, 2019). This arises largely as environmentalists challenge the development status quo, seeking the prevention or minimisation of large infrastructure projects, advocating for the maintenance of nature areas, pollution controls, and so on, in order to curb the nepotistic and non-transparent ways in which mafia-states further the interests (and personal fortunes) of the few. Environmentalists also often pretend to hold democratic ideals such as transparency in decision-making and planning, insisting that citizens have a place at the table with decisions that impact them.

We proceed below with attention to three kinds of power: first, interpersonal, second, institutional (the silencing and hollowing out of environmental expertise from within the civil service and key governmental institutions); third, societal-structural (outright attack and propaganda against civil society). All three examples demonstrate how the Orbán regime positions and values loyalty to the party line above public interest. Our consideration of the targeted attacks on a single individual as part of the demise of environmental research at St István University (within this the KTI Institute) shows how the intentions behind key decisions were not necessarily the decimation of a part of the environmental sector, but rather the elimination and ruination of a previously successful individual, with a wider institution or public being mere collateral damage. Our consideration of the silent dismissal of environmental expertise from public institutions reveals how the Orbán regime has eliminated formal environmental consultation.

Outright, aggressive attacks on the civil sector demonstrate the regime's willingness and ability to use state-funded propaganda, law and violent arms of the state to intimidate and repress. Furthermore, a reckoning with the lack of broad societal protest or response to the events described below is key to understanding how authoritarianism has emerged and been consolidated in Hungary. We will return to some consideration of (the lack of) societal responses to the events we describe in our Discussion.

Embedded in the Collapse

Before commencing our review of the environmental sector's decline in Hungary, a word about our own positionalities, information access and sources. For the time period covered, we have both been university-based researchers, with active fieldwork engagement, sites, and in one case, daily life in a rural area of Hungary. As individuals with personal passions for conservation and the environment, our research interests have intersected directly with the experiences and changes the environmental sector has undergone over the past decade. Alongside meeting and interviewing, independently, hundreds of farmers throughout the country, we have each engaged, in different capacities, with informal and formal work consultancies and opportunities through the different incarnations of the Environment Ministry and various national parks, and we have hobbies that are organised through civil volunteer organisations and national park directorates. One of us was also deeply involved in ecopolitical civic activism from the 1990s until 2009. For full disclosure, we also both worked as part of a small research team (one formally and one in an affiliated capacity) at the KTI Institute that is the subject of analysis below.

Due to these personal histories, in many cases we and our close acquaintances have lived through, witnessed and directly experienced the decisions and consequences of government closure, dismissal, de-financing, and de-legitimation. Our access and histories also enable us to compare—from our own memories and experiences, but also through those of a number of generous interlocutors—the aspirations and environmental work of a decade ago, to that of today. For the purposes of this chapter, we draw on a number of conversations and interviews with colleagues at KTI, formerly (and currently) of national park directorates and ministries, and with individuals from within these fields with whom we have continued personal friendships. We have complemented assertions made regarding budget cuts and ministerial decisions with external sources, drawing from widely circulated and publicly available governmental speeches, blogs and press coverage to piece together what may at first appear to be a fragmented story.

Interpersonal Workings of Authoritarian Power: The Demise of KTI

During the summer of 2014, media reports filtered through the following decision released by the St István University Senate under the title "rationalisation":

> There is ongoing organisational restructuring [...] that aims to produce organisational units of a professionally clean profile. Within this restructuring, the scope of duties of the six departments of KTI will be reorganised into two institutes [...] which enables a more rational and manageable operation from a management and governance perspective (SzIE, 2014)

This official text, appealing to rational management and efficient governability, signalled one of the most irrational decisions made by a Hungarian university in the post-transition period. The decision of university administration to "restructure" and thus effectively eliminate a successful academic unit cannot be understood through common higher education measures of success. The call to rationalisation made invisible and diminished KTI's academic and professional achievements. At the time of the announcement, KTI was the only institute of the Faculty of Agricultural and Environmental Sciences that contributed a positive financial balance to the faculty budget. Researchers at KTI brought in more EU-funded projects than all other institutes of the university together. Two education programmes run by KTI (agri-environment and nature conservation engineering) were the most popular bachelor's and master's qualifications in terms of applications; and KTI faculty served as policy advisors to public administration at multiple levels in Hungary. The decision to close the institute and the way in which this closure occurred can only be understood with reference to the interpersonal relations between the institute's director and Prime Minister Viktor Orbán, and the latter's prerogative of maintaining *Fidesz* party loyalty.

KTI was formed in 1990 as an interdisciplinary institute at the intersection of agriculture, rural development and sustainability studies, singular in the country for its policy-oriented research and interdisciplinary education and commitment to rural revival that aligned nature conservation with small-scale sustainable agriculture. The remit of the institute was unique in the Hungarian—and arguably regional—context, as the institute aimed to train not agricultural technical specialists, but people who could think about the countryside holistically. This holism was clearly manifested in the agri-environmental engineering education programme, which attempted interdisciplinarity through integrating ecological, biological, engineering, environmental, social scientific and humanities approaches to farming and rural development.

KTI's director from the mid-1990s was Professor József Ángyán, who had a reputation not only academically, but as a prominent advocate for farmers' rights and rural sustainability. Ángyán was a widely recognised national figure. He personally spearheaded public-policy programmes for rural development that focused on the interests of small-scale, less commercially oriented family farmers. Further, he had a widely documented role in organising successful farmer protests in 2005 against the then Gyurcsány government's handling of new EU agricultural payments (direct EU support for farmers commenced after Hungary's accession to the EU in 2004). As two rural field researchers ourselves, we personally experienced the wider recognition and esteem for Ángyán's name during our own fieldwork, where even in the most remote and far-flung places we were asked, "Do you know Professor Ángyán?!"

Ángyán's Christian-democratic ideological stance and recognised commitment to the agricultural community were crucial to the political uptake of KTI's academic ideas. His renown led to his being approached directly by Orbán and *Fidesz* representatives to stand within the *Fidesz* faction for the 2010 parliamentary elections (he had been an independent MP since 2006), and to give his name and rural development strategy to *Fidesz* as its own policy. The hope was that Ángyán's standing within a crucial demographic of rural voters would provide an aura of authenticity to *Fidesz* in the countryside. After Orbán's win, Ángyán was appointed Under-Secretary to the Minister for Agriculture and Rural Development, where Ángyán hoped that his appointment would give rise to an unprecedented opportunity for collaboration between agriculture, nature conservation and rural development in public administration.[2]

Such high hopes receded as it became clear that there were enormous discrepancies between Ángyán's rural development vision (which was available on the ministry's website until early 2019) and the actual operation and realisation of its agenda. From 2011, EU agricultural and rural development funding was redirected towards agribusinesses. Land policy also presented an opportunity for elite capture, as twenty-year government leases expired and required renewal at this time. In contrast to Ángyán's public statements about making this and other land available to family small-holders, a high proportion of state lands were first leased and then sold to those with demonstrable ties to the political elite, and who were in fact neither local, nor farmers (Ángyán, 2018, 2016; Gonda, 2019).

As an MP Ángyán contested and highlighted this divergence. According to interviews and personal accounts provided since, this scrutiny was regarded by Orbán as an act of 'betrayal' of *Fidesz*, for which Ángyán experienced real consequences. Soon after Ángyán's public resignation, the communications about the KTI restructuring and rationalisation began within the university. Over a single summer, legislative policy was passed by the Hungarian Parliament whereby all higher education institutes were deprived by law of the autonomy to choose their own academic (rector) and economic (later called chancellor) managers. These positions were subsequently appointed by the government: the Dean of St István University immediately initiated "organisational restructuring" that led to the "professionally clean profile" whereby KTI was closed down, and one man's career achievement was erased. In addition to KTI's closure, the Ángyán-developed 'agri-environmental engineering studies' was removed by government decree as a recognised bachelor's diploma from the official

2 Ángyán has now given a series of public interviews about this time and its promises, for example, see Nagy, 2020.

list of higher education programmes. Over the following two years, KTI workers were gradually released from contract and disbanded to different departments and institutes, while others resigned and left St István University as an act of protest.

KTI's closure may be read as a highly individual, personalised story of revenge against a single person, as Professor Ángyán broke ranks and attempted to draw widespread public attention to *Fidesz*'s rural development policy and nepotistic land allocations. The decimation of Ángyán's achievements served as an example and warning to continuing *Fidesz* party faithfuls, many of whom were also duly rewarded for their parts. For example, Csaba Gyuricza, a former student of Prof. Ángyán, became the University's new dean in 2014. Soon after Gyuricza oversaw the closure of the institute in which he had studied, he was next appointed to the directorship of the National Agricultural Research and Innovation Centre (today he is the appointed Rector responsible for the restructuring of the now renamed St István University—from January 2021 known as the Hungarian University of Agricultural and Life Sciences).

The lack of personal solidarities or public protest in response to the undermining of a respected individual and of a successful, internationally recognised institute are also integral parts of this narrative: fellow parliamentarian ministers admitted to not having the "freedom" to speak in Ángyán's support (Nagy, 2020); critical accounts of KTI's closure only appeared in independent blogs and not the wider media (see e.g. Greenfo, 2014). This speaks to the systemic, ingrained nature of these power plays and decisions. KTI's closure was significant in Hungarian educational and research life, as the institute had been the most successful of its kind in the country: the place that farmers' kids went to train. In turn, these events merely preceded similar actions by the Orbán regime to institutionally re-develop (and effectively privatise) the wider higher education and academic sector, a process that is ongoing today (see the fate of the Hungarian Academy of Sciences, and—as of January 2021—almost all public universities in Hungary).

Thus within three years of Orbán's re-election, state power was informed by personal enmity and vengeance rather than any 'rational' decision-making process. Within *Fidesz* party meetings, Orbán let it be known that Ángyán would be "punished": Orbán did not publicly fire nor even reprimand Ángyán, as he allegedly told Ángyán that, "we will not make a martyr out of you," (personal communication, see also Nagy, 2020). This is the realisation of Magyar's mafia-state theory against 'traitors' who fail to adhere to the party line, and who must be punished: Ángyán's punishment was to see the institutional decimation of his life's work in education. The concept of the mafia-state transposes the internal logics of mafia networks, where fealty to the 'family' takes precedence above all else and is enforced. The ease with which university institutions and degree programmes were re-ordered administratively and bureaucratically is a result of the absolute majority the Orbán government enjoys in parliament, which allows them to rule by decree. This case also demonstrates the extensive reach, personal connections and ability of this government to transpose and allocate new roles to party acolytes willing to undertake actions deemed necessary by superiors.

Institutional Take-over: Dismissal of Experts from within Public Administration

This section summarises and documents the ever-more direct elimination of environmental expertise and evidence-based decision-making systems from within formal state environmental institutions in Hungary, such as its ministries and the country's national park directorates. It should, however, be emphasised that the weakening of the actual implementation capacities of environmental regulation had started much earlier than the second Orbán government. Previous analyses have shown that several institutional adjustments stemming from the environment-related legal requirements of the EU accession process resulted in higher workloads and fewer available resources for environmental street-level bureaucrats. In particular, field-level environmental activities, including site controls, had previously lost out owing to institutional restructuring (Jávor and Németh, 2008; 2007). The ruling socialist-liberal coalition governments of the early 2000s gave clear policy priority to economic growth and investors' interests over environmental values and community needs (e.g. by loosening the control capacity of environmental regulation over large infrastructural investments, declaring them to be of national interest, etc.).

From 2010 onwards, the second Orbán government aimed to trim the number of people working within the civil service. In practice these changes were carried out at great cost to the environment sector. In 2010, the Ministry for Environment and Water was closed, and responsibility for water-related public administration was transferred to the Ministry for Internal Affairs. The competency and power of environment and nature conservation efforts were seriously reduced as they were transferred to the Ministry for Rural Development (currently the Ministry for Agriculture). The political representation of nature conservation suffered the most: since these changes, "there has been no separate deputy state secretary position for nature conservation" in the Hungarian government (Mihók et al., 2017: 70).

Regional environmental agencies came to the same fate. The Hungarian environmental street-level bureaucracy consisted of a regional system of environmental state authorities and national park directorates. The former exercised power to monitor industrial plants and factories and to initiate on-site inspections. There are ten national park directorates organised regionally across the country, that used to hold state legal authority (implementation power) for nature conservation. In 2015, regional environmental agencies were deprived of their organisational autonomy for implementation and decision-making as a result of government restructuring and were integrated into county-level government offices at a department level. Essentially, these changes meant that local oversight was eliminated, with responsibility for any local issues placed within more 'macro' institutions that were also responsible for other issues.[3] This constituted a clear downgrading of regulatory power of the environmental sector and has made it easier for governments to overrule advice and assessment regarding the environment and nature conservation. County office managers are today politically appointed, often without environmental qualifications, expertise or experience, in contradistinction to earlier times. They

3 Prior to 2015, governmental environment offices worked at the regional level, which was the same administrative unit as the national parks. With the elimination of regional offices, only the county-level oversight offices remain. Importantly, environmental expertise within national parks has remained at the regional level, meaning that coordination has become difficult, with county offices not solely responsible for environmental oversight, meaning these issues have become watered down, and with overall fewer people working on these issues.

frequently override the professional environmental opinions of their lower-ranked colleagues. These changes have not been remarked upon by the European Commission and other EU institutions.

In addition to these losses in regulatory oversight powers, the Hungarian government has financially starved the environmental sector. In 2012, the ombudsman for constitutional rights commissioned a study on the impacts of economic-crisis management by the Orbán government. The report concluded that, among other vulnerable and marginalised social groups (e.g. the homeless) and social sectors (e.g. health and social policy), a major economic loser had been the environmental sector (Szajbély, 2013). The dramatic decline in budgetary support can be demonstrated through national park directorates: in 2008 government finance amounted to 52.2% of park expenses; in 2011, the same figure was 15.9% (Kovács, 2017). Conservation projects today exist almost entirely through European Commission grants, specifically from LIFE, LIFE+, and less frequently Horizon 2020 financing (ibid.). Centralisation has also increased: all revenues from environmental charges and fees have been redirected to the central government budget instead of being earmarked and used by state environmental agencies, as was previously the case.

There have been sizeable staff lay-offs between 2012 and 2019: in 2012, 25% of staff were made redundant from the environmental undersecretariat of the Ministry for Rural Development relative to 2011; in the autumn of 2018, in the name of "reducing unnecessary bureaucracy", 44% of nature conservation experts at the Ministry of Agriculture lost their jobs (this was characterised as a "bloodbath" by the few independent blogs that covered the story).[4] The most recent dismissals have been devastating to ongoing projects and the continuity of programmes. The immediate outcome has been a skeleton staff working on projects, and a lack of environmental expertise to draw from in public policy decisions. For example, the Ministry of Agriculture no longer has a Natura 2000 taskforce (as a result of contract termination and resignation), such that these areas no longer have a management group at the national level, and local managers are unclear as to their point of contact. The concrete

4 https://index.hu/belfold/2018/11/09/lefejeztek_a_termeszetvedelmet/; https://qubit.hu/2018/12/11/latvanyosan-kivereztetik-a-nemzetkozi-hiru-magyar-termeszetvedelmet.

reasons for staff dismissal from these areas are open to conjecture. The Orbán government has frequently stated that its civil sector is bloated, but the environment sector has been disproportionately cut. This has had significant consequences for the systems of oversight relating to nature areas and land users, and has led to the diminishment of environmental protection voices from the wider public arena.

These patterns within the Ministry in Budapest are replicated across the country. As the Hungarian government sought a sleeker civil service, numerous hiring freezes were instigated over the past five years, such that governmental agencies have often been unable to hire individuals to work on even externally-funded projects. As the majority of conservation initiatives today are externally funded, undertaken by national parks, the parks' inabilities to employ staff mean that they are unable to complete these projects. In 2010, the second Orbán government changed the directorship of a number of national parks, instating in leadership roles individuals without backgrounds or training in conservation. Over time, leadership changes from the top gradually gave rise to changes throughout regional offices. Several office managers describe a greater push from ministerial levels to make a profit from national-park managed land, in particular from forests and agricultural areas. Consequently, national parks have heavily invested in agricultural mechanisation and livestock on their own territories, effectively instigating competition with surrounding land users. National parks were formally recognised as agricultural land users in the updated 2020 land law, which designated them "agriculture producer organisations".[5]

Biodiversity and mapping assessments and production plans for arable lands and forests are no longer circulated in-house within national parks, due to perceptions from management that biologists' opinions on plans are 'obstructionist'. As a consequence, the ratio of technical staff (i.e. agricultural engineers, foresters) has grown relative to those with biology or ecological skills. Formerly prevalent decision-making pathways *within* national park offices have been sidelined, if not formally scrapped.

5 2020 XL law on the settlement of land ownership of producer cooperatives plots and amending certain laws on land matters, see https://net.jogtar.hu/jogszabaly?docid=A2000040.TV#lbj0id5e29.

The dismantling of the governmental environmental sector through dismissal, financial starvation and dilution of expert opinion and decisions has widespread practical consequences. These include inadequate or negligent management of conservation areas, and the greenlighting of formerly protected areas for the benefit of private or large infrastructure projects. Many former workers in the environment sector have moved into non-governmental or entirely unrelated roles, or taken early retirement, suggesting a significant brain-drain of expertise and commitment.

The weakening of the sector has been the work of successive governments. While the primacy of business interests over environmental ones, corruption and non-transparent decision-making pathways have been features of all governments (whether conservative, liberal or socialist) from the 1990s onwards, the period after 2010 has seen an acceleration and normalisation of these trends: "a stronger fusion of economic and state power" (Scheiring and Szombati, 2020: 727). Cuts to the civil service have also effectively streamlined numerous development proposals of the "accumulative state" (Scheiring, 2019) that were previously caveated or obstructed by environmental considerations. The concept of the accumulative state describes Hungary's developmentalist path since 2010 as a new alliance between domestic and foreign capital and the political elite. Orbán has effectively chosen an authoritarian strategy to intensify the embeddedness of the Hungarian economy in the neoliberal operation of global value chains dominated by transnational companies (Scheiring and Szombati, 2020). The authoritarian re-structuring of governance institutions ("the foundations of Hungary's polity, society and economy", Scheiring and Szombati, 2020: 722) led to the gradual dismantling of independent state bureaucracy and its associated expertise (including all levels of environmental administration), democratic checks and balances. This was accomplished through the purchase of the majority of available media, with the active assistance of domestic and foreign investors, which enabled an effective governmental propaganda machine to take root (such that alternative or dissident voices are not aired or heard). Orbán's parliamentary super-majority has led to the government's rewriting of the Hungarian constitution into a modifiable, so-called "fundamental law", which no longer contains effective protections for the rights of citizens to a healthy environment. A further prong in this authoritarian consolidation was the limitation of the independence of courts, which has had dire implications for the enforcement of community and citizen environmental interests through the law (see Pech and Scheppele, 2017 for an overview of these processes).

The accumulative state builds smoothly upon a mantra of economic growth and transnational capital interests, which are hindered by a strong environmental sector (as well as environmental expertise within state bureaucracy and civil society organisations). In the same vein, all direct democratic institutions, and those that lobby for citizen participation, transparency and control over decision-making, pose the same hindrance to the accumulative interests of the state and transnational capital. Below, we outline the current status of the civil society sector and the aggressive displays of power it has endured in recent times.

Attacks and Silencing: The Civil Society Sector

In many countries, the most articulate and passionate environmental campaigners are found in the non-governmental and civil sector. Under emergent authoritarian regimes, the legitimacy of civil society organisations is questioned as they are demarcated as foreign and unelected and thus 'undemocratic'.

With hindsight, the 1980s environmental movement held promise that never materialised in the region. During the 1990s, globalisation saw the establishment, expansion and influence of international conservation organisations. These agencies are typically headquartered overseas, are better-financed than local, organic outfits, and have brought in conservation projects and financing forms, with the expectation that local conservation actors would adopt and play catch-up. This model of development and mimicry, termed "projectification" by Adrian Swain (2007), is one where development and conservation operate only project-to-project and are financed almost entirely by external funding sources. Throughout eastern Europe this sort of projectification has appeared in the form of the European Commission's LIFE projects, or through the CAP's rural development or Leader Programme.

The Orbán government applies a political rhetoric that questions the existing NGOs' representative nature within Hungarian society. This approach has extended to accusing groups financed by foreign foundations of meddling in domestic affairs by importing foreign expectations and norms. The best example of how such accusations have played out is the fate of the Norwegian Foundation (NF). During late 2014 (as KTI was also 'rationalised'), after months of public statements from the then deputy PM János Lázár questioning the integrity and mission of the NF, the police-led National Bureau of Investigation raided the offices of NGOs that were the recipients of financing from the NF, such as *Ökotárs* (the Environmental Partnership Foundation). The raids were conducted to intimidate: inspectors and detectives used force to access offices, and raided the private homes of NGO directors (actions that were later found to be illegal and a breach of due process, as there was no accusation of criminal activity). The allegations against the NF (and its selected in-country representatives, such as *Ökotárs*) from the Orbán government were that it only financed Hungarian organisations that were politically aligned with its left-leaning, progressive ideals. These outfits acted in ways oppositional to the ruling government, and as such, the NF allocated money in a biased, subjective way. Second, recipients were accused of spending their grants non-transparently and failing to follow due procedure with their funds.

Due legal inquiry cleared the NF and NGOs of wrong-doing, and these actions were never about criminal liability. However, the scandal set the stage for the next phase of governmental attack. First, these organisations' names were besmirched and widely questioned by a partisan media. Political criticisms of NGOs' lack of embeddedness within society are somewhat founded, and several environmental NGOs do struggle to demonstrate local support for their conservation messages. In this context, environmentalists can be relatively easily delegitimised through government-controlled press and spokespeople, through an ideological positioning of them as working for the 'Other'—and thus by extension being an 'Other'. They are branded as not operating in the interests of the nation or Hungarian civil society, but as bought out (and brought in) by foreign interests.

Second, claims that NGOs operate under a democratic deficit were a core component of Orbán's 'illiberal democracy' outlined in late

July 2014. A new system of government oversight into foreign funds to NGOs was passed by the Hungarian Parliament in 2017 (*Lex NGO*, Act no. LXXVI): this legal package requires all CSOs to publicly brand and register their sources of finance, even from individuals, and to state that they are recipients of foreign financing on any correspondence, advertising or marketing. These rules apply compulsorily to NGOs working in the human rights and environmental fields, while other sectors (such as business or private companies, or those financed by the Hungarian government), are exempt. The law also created a public list of all organisations registered under this act.[6]

An important feature of the outright aggressive disciplining of the civil society sector is that the Orbán-led government controls definitions and narratives, determining who is viewed as 'in' or 'out'. Today, the government has gradually co-opted civil society by financing organisations that are its allies, such that an "uncivil" illiberal society has emerged (see Molnár, 2020). The lack of local connections and societal buy-in for most NGOs has facilitated the increased use of repressive tactics such as those documented above, and the expanding reach of Orbán's rightist and Magyar-centric ideology. The aggressive machinations of intimidation and closure were trialled and executed by the Orbán government initially on the environmental sector, and subsequently extended to any other sphere that may hold independent or critical (and thus potentially anti-government) collectives (see for example the current machinations around the higher education sector).

Discussion

The environmental sector experienced a difficult decade in Hungary between 2010 and 2020. It has been subject to political attack and greatly

6 In February 2021, the European Court of Justice initiated further proceedings against the Hungarian government for failing to address its earlier June 2020 judgement, which found that *Lex NGO* breached fundamental freedoms contained in the Charter of Fundamental Rights of the European Union, namely, the free movement of capital, right to privacy, and freedom to associate. See http://civil.info.hu/kulfoldrol-tamogatott-civil-szervezetek and https://m.hvg.hu/eurologus/20210218_jogallamisag_Magyarorszag_europai_birosag_kotelezettsegszeges?fbclid=IwAR2QlAz6SedStqMb_mDY2QznJIyTcJFnfyz6zrDE9_Yjxf4olFdc9_NH9AY.

reduced state financial (and ideological) support, mass redundancies and a hollowing out of research and governmental institutes, which has also resulted in the dismantling or dilution of decision-making pathways within environmental agencies.

In this context, large infrastructural and development projects requiring environmental assessments become mere rubber-stamp additions to plans. It is important to recognise that *Fidesz* has only built on and expanded the legal inventions of the previous socialist-liberal coalition government, which simplified and sped up investments deemed to be of national economic significance (see Act no. LIII of 2006).[7] Underpinning the success of the Orbán regime's moves against the environmental sector is its communications strategy, achieved through a media coup of TV, radio, newspapers and increasingly even online media, via the purchase of these outlets by domestic and foreign *Fidesz*-friendly entrepreneurs and capitalists. In addition, key government figures frequently speak mockingly of environmental issues, belittling the perceived "urgency" of climate change relative to other changes they deem more of a threat, such as immigration (as for instance uttered by the Speaker of the House, László Kövér; see Simon, 2020). These derogations underscore the weakness of the environmental sector, which due to its inherent fragmentation, lack of social embeddedness and media access, struggles to respond or gain ground.

Societal responses have also been muted or absent. A notable exception were the responses to *Lex NGO* in 2017, when an NGO umbrella organisation launched petitions and a number of street demonstrations, appealing to Europe and Europeans for solidarity.[8] On the whole, however, these protests remained localised in urban centres. While the Orbán government seeks to diminish the voice, reach and strength of the environmental community, it is also crucially able to do so because of the current economic and social divisions and state

7 This regulation explicitly referred to the effective implementation of EU-funded projects, but has in practice contributed to reductions in transparency and accountability for all government-supported development projects.

8 https://www.greenpeace.org/hungary/sajtokozlemeny/2440/75-kornyezet-es-termeszetvedo-civil-szervezet-allasfoglalasa-a-kulfoldrol-tamogatott-szervezetek-atlathatosagarol-szolo-torvenyjavaslatrol/.

of Hungarian society, wherein environmentalism is not a widespread, shared concern.

Political attacks on the sector serve not only to sideline environmental messages, but also to socially fragment and silence parts of society that are viewed as oppositional to government. Political ecology conceives of social, economic and environmental alternatives as having "their own bases in power complexes situated in social movements, trade unions or the forces of civil society, and are distinctive in their use of more informal media of thought, discussion and dissemination" (Watts and Peet, 2004: xiv). Clamping down on these alternatives and their methods of communication; questioning and undermining their very legitimacy removes the option for the promulgation, let alone adoption, of alternatives. The lack of development of the environmental sector is an outcome of the deliberate undermining of the civil society sector as a whole, and thus represents a telling case study for 'illiberal democracy' realised.

Achieving an 'illiberal democracy' in the Hungarian context requires the redirection of the country's economy and development trajectory down a deeply neoliberal pathway as well as clamping down on any third sector that jeopardises or effectively criticises that pathway. The dismantling of the environment sector has had broad socio-political and environmental consequences. Ineffective government agencies today rubber-stamp decisions and deflect attention from otherwise valid environmental considerations, leading to a reduction in environmental standards and management, and the degradation of conservation areas. Natural resources are largely 'up for grabs' as established systems of oversight (and even European-level protection, as with Natura 2000 areas) no longer have the capacity to enforce rules (or are granted outright permission to build on protected Natura 2000 wetland, as in the case of an Audi factory near Győr). The re-allocation of resources and industries is also in full swing—see the land issue over which Ángyán resigned from his post—with the rules written and modified to suit the access of local elites.

Our documentation of the progressive weakening and effacement of environmental expertise in higher education, public administration, and the civil or third sector in this chapter makes clear the authoritarian state's obsession with all-encompassing power. The story of KTI in particular underscores the naivety of expectations that positive societal visions or values are required for the operation and realisation of governmental plans. The seemingly interpersonal clash between Ángyán and Orbán includes instead a general message and, perhaps more importantly, a warning to those inside the ruling elite about the costs and risks of speaking out, as compared to strategies of party and personal loyalty.

The deliberate dismantling of environmental public administration demonstrates the same synergies between the interests of an authoritarian state and global corporate actors. This synergy is not at all unique to the Hungarian case and, most probably, provides an answer as to why this dismantling has gone virtually unnoticed within the European Union. Despite the authoritarian turn and some negative attention and rhetoric around Hungary's 'rule of law failures' from the EU, there is at the same time a strategic alliance between the Orbán regime and European and other multinational corporations. In the current context, where environmental administrations are resource-poor and politically controlled, *any* environmental commitments by these global market players may be deemed remarkable: the underlying dynamics of the "accumulative state" is one primarily interested in "having more" and "growing bigger and faster", as opposed to any concerns for the general well-being of citizens (Scheiring, 2019).

The neoliberalised Hungarian context is also evident if we look at the struggles of local communities. The power of capital, legally backed by the Orbán regime, has overwhelmed the playing field against the rights of citizens for a healthy environment. Contemporary environmental struggles tend to mirror those of the socialist-authoritarian era of the 1970s and '80s: local residents try to bring forward their own well-being interests against the profit accumulation interests of the alliance between state and big capital. The politically threatened and resource-weakened ENGO sector cannot be of much help to these local environmental struggles. Apolitical tendencies have reemerged amongst Hungarian ENGOs, instead of explicitly political ecological commitments. Therefore, the dismantling of environmentalism in Hungary shows that authoritarian political institutional arrangements may well serve both the accumulative interests of global market actors (private corporations) and the state.

References

Ángyán, J., 2018. II. 'Megyei elemzések (Zárójelentések az állami földprivatizációs rendszer valós értékeléséhez) Borsod- Abauj — Zemplen megye'. https://greenfo.hu/wp-content/uploads/dokumentumtar/allami-foldprivatizacio--intezmenyesitett-foldrablas-borsod-abauj-zemplen-megye.pdf.

——, 2015. 'Állami földprivatizáció — intézményesített földrablás'. https://alfahir.hu/sites/default/files/kepek/allamifoldprivatizacio.pdf.

Bogaards, M., 2009. How to classify hybrid regimes? Defective democracy and electoral authoritarianism. *Democratization*, 16, 399–423. https://doi.org/10.1080/13510340902777800.

Buzogány, Á., 2015. Representation and Participation in Movements. Strategies of Environmental Civil Society Organizations in Hungary. *Südosteuropa. Zeitschrift für Politik und Gesellschaft*, 63(3), 491–514. https://www.academia.edu/18283908/Representation_and_Participation_in_Movements_Strategies_of_Environmental_Civil_Society_Organizations_in_Hungary_S%C3%BCdosteuropa_63_2015_no_3_pp_491_514.

Cent, J., Mertens, C., Niedziałkowski, K., 2013. Roles and impacts of non-governmental organizations in Natura 2000 implementation in Hungary and Poland. *Environmental Conservation*, 40, 119–28. https://doi.org/10.1017/S0376892912000380.

Corry, O., 2014. The Green Legacy of 1989: Revolutions, Environmentalism and the Global Age. *Political Studies*, 62, 309–25. https://doi.org/10.1111/1467-9248.12034.

Cox, T., 2020. Between East and West: Government–Nonprofit Relations in Welfare Provision in Post-Socialist Central Europe. *Nonprofit and Voluntary Sector Quarterly*, 49, 1276–92. https://doi.org/10.1177/0899764020927459.

Faragó, K., Vári, A., Vecsenyi, J., 1989. Not in My Town: Conflicting Views on the Siting of a Hazardous Waste Incinerator. *Risk Analysis*, 9, 463–71. https://doi.org/10.1111/j.1539-6924.1989.tb01257.x.

Gonda, N., 2019. Land grabbing and the making of an authoritarian populist regime in Hungary. *The Journal of Peasant Studies*, 46, 606–25. https://doi.org/10.1080/03066150.2019.1584190.

Gottfried, S., 2019. Oligarchs, Oligarchy, and Oligarchization, in: Gottfried, S. (Ed.), *Contemporary Oligarchies in Developed Democracies*. Springer International Publishing, Cham, pp. 21–62. https://doi.org/10.1007/978-3-030-14105-9_2.

Greenfo, 2014. Pár tucatnyian temették Ángyán József gödöllői intézetét [WWW Document]. *Greenfo*. https://greenfo.hu/hir/par-tucatnyian-temettek-angyan-jozsef-godolloi-intezetet/.

Haraszti, M., 1990. The Beginnings of Civil Society: The Independent Peace Movement and the Danube Movement in Hungary, in: Tismaneanu, V.

(Ed.), *In Search of Civil Society : Independent Peace Movements in the Soviet Bloc*. Routledge, New York, pp. 71–87.

Harper, K., 2006. *Wild Capitalism: Environmental Activism and Postsocialist Political Ecology in Hungary*. East European Monographs, University of Massachusetts Amherst, MA.

Innes, A., 2014. The Political Economy of State Capture in Central Europe. *JCMS: Journal of Common Market Studies*, 52, 88–104. https://doi.org/10.1111/jcms.12079.

Jávor B., Németh K., 2008. Reformok, megszorítások és a környezetvédelem hatósági rendszere. *Politikatudományi Szemle*, 17, 35–60. https://poltudszemle.tk.hu/szamok/2008_3szam/2008_3_javor.pdf.

——, 2007. Kisebb állam, nagyobb baj?: Beszámoló a zöldhatósági rendszer kialakításának értékeléséről. *Társadalomkutatás*, 25, 487–511. https://doi.org/10.1556/tarskut.25.2007.4.6.

Kerényi, S., Szabó, M., 2006. Transnational influences on patterns of mobilisation within environmental movements in Hungary. *Environmental Politics*, 15, 803–20. https://doi.org/10.1080/09644010600937249.

Kovács, E., 2017. A nemzeti park igazgatóságok 2000–2015 közötti költségvetésének értékelése az alapfeladataik tükrében (Assessment of the Hungarian national park directorates' budget between 2000 and 2015 in the light of their main tasks). *Természetvédelmi Közlemények*, 23, 201–23. https://doi.org/10.17779/tvk-jnatconserv.2017.23.201.

Magyar B., 2013. *Magyar polip: A posztkommunista maffiaállam*. Noran Libro Kiadó, Budapest.

Mawdsley, E., Mehra, D., Beazley, K., 2009. Nature Lovers, Picnickers and Bourgeois Environmentalism. *Economic and Political Weekly*, 44, 49–59.

McCarthy, J., 2019. Authoritarianism, Populism, and the Environment: Comparative Experiences, Insights, and Perspectives. *Annals of the American Association of Geographers*, 0, 1–13. https://doi.org/10.1080/24694452.2018.1554393.

Mertens, C., 2013. Playing at multiple levels in biodiversity governance: The case of Hungarian ENGOs in Natura 2000. *Society and Economy*, 35, 187–208. https://doi.org/10.1556/socec.2013.0001.

Mihók, B., Biró, M., Molnár, Z., Kovács, E., Bölöni, J., Erős, T., Standovár, T., Török, P., Csorba, G., Margóczi, K., Báldi, A., 2017. Biodiversity on the waves of history: Conservation in a changing social and institutional environment in Hungary, a post-soviet EU member state. *Biological Conservation*, 211, 67–75. https://doi.org/10.1016/j.biocon.2017.05.005.

Mikecz, D., 2017. Environmentalism and civil activism in Hungary, in: Moskalewicz, M., Przybylski, W. (Eds.), *Understanding Central Europe*. Routledge, London, pp. 359–65.

Molnár, V., 2020. Civil society in an illiberal democracy: Government-friendly NGOs, "Foreign Agents" and Uncivil Publics, in: Kovács, J. M., Trencsényi, B. (Eds.), *Brave New Hungary: Mapping the "System of National Cooperation"*. Lexington Books, Lanham, Boulder, New York and London, pp. 51–72.

Nagy, J., 2020. Ángyán József: Viktor azt mondta, "addig maradsz, ameddig kibírod, nem fogunk mártírt csinálni belőled." *24.hu*. https://24.hu/belfold/2020/05/19/angyan-jozsef-rendszervaltas30-orban-fidesz-fold-interju/.

Nyyssönen, H., Metsälä, J., 2021. Liberal Democracy and its Current Illiberal Critique: The Emperor's New Clothes? *Europe-Asia Studies*, 73, 273–90. https://doi.org/10.1080/09668136.2020.1815654.

Pavlínek, P., Pickles, J., 2000. *Environmental Transitions: Transformation and Ecological Defense in Central and Eastern Europe*, 1 edition. Routledge, London and New York.

Pech, L., Scheppele, K. L., 2017. Illiberalism Within: Rule of Law Backsliding in the EU. *Cambridge Yearbook of European Legal Studies*, 19, 3–47. https://doi.org/10.1017/cel.2017.9.

Pető, A., Grzebalska, W., 2016. Around the Bloc: How Hungary and Poland Have Silenced Women and Stifled Human Rights. *Transitions Online*, 10/25, 28–31.

Polese, A., 2014. Informal Payments in Ukrainian Hospitals: On the Boundary between Informal Payments, Gifts, and Bribes. *Anthropological Forum*, 24, 381–95. https://doi.org/10.1080/00664677.2014.953445.

Polese, A., Morris, J., Kovács, B., 2015. Introduction: The Failure and Future of the Welfare State in Post-socialism. *Journal of Eurasian Studies*, 6, 1–5. https://doi.org/10.1016/j.euras.2014.11.001.

Reynolds, D. A. J., 2020. Let the River Flow: Fighting a Dam in Communist Hungary. *Hungarian Cultural Studies*, 13, 111–30. https://doi.org/10.5195/ahea.2020.391.

Sachs, W., 1995. Global ecology and the shadow of development, in: Sachs, W. (Ed.), *Global Ecology*. Zed Books, London, pp. 3–21.

Scheiring, G., 2019. Dependent development and authoritarian state capitalism: Democratic backsliding and the rise of the accumulative state in Hungary. *Geoforum*. https://doi.org/10.1016/j.geoforum.2019.08.011.

Scheiring, G., Szombati, K., 2020. From neoliberal disembedding to authoritarian re-embedding: The making of illiberal hegemony in Hungary. *International Sociology*, 35, 721–38. https://doi.org/10.1177/0268580920930591.

Simon K. B., 2020. Orbán harca a környezetvédelem ellen. *Magyarnarancs.hu*. https://magyarnarancs.hu/belpol/a-globalizacio-fedoszerve-125247.

Snajdr, E., 2008. *Nature Protests: The End of Ecology in Slovakia*. University of Washington Press, Seattle, WA and London.

Stepurko, T., Pavlova, M., Gryga, I., Murauskiene, L., Groot, W., 2015. Informal payments for health care services: The case of Lithuania, Poland and Ukraine. *Journal of Eurasian Studies*, 6, 46–58. https://doi.org/10.1016/j.euras.2014.11.002.

Swain, A., 2007. Projecting "transition" in the Ukrainian Donbas: Policy transfer and the restructuring of the coal industry, in: Swain, A. (Ed.), *Re-Constructing the Post-Soviet Industrial Region: The Donbas in Transition*. Routledge, London, pp. 161–87.

Szabó, M., 2000. External Help and the Transformation of Civic Activism in Hungary. *Javnost — The Public*, 7, 55–70. https://doi.org/10.1080/13183222.2000.11008735.

Szajbély, K., 2013. A válság vesztesei — a paragrafusok fogságában (AJBH Projektfüzetek No. 2013/3). https://www.parlament.hu/irom39/10126/pdf/valsag_vesztesei.pdf.

Szalai, J., Svensson, S., 2018. On Civil Society and the Social Economy in Hungary. Intersections. *East European Journal of Society and Politics*, 4, 107–24. https://doi.org/10.17356/ieejsp.v4i4.471.

SzIE, 2014. A Környezet- és Tájgazdálkodási Intézet racionalizálásáról | SZIE [WWW Document]. https://szie.hu/kornyezet-es-tajgazdalkodasi-intezet-racionalizalasarol.

Tóth, C., 2014. Full text of Viktor Orbán's speech at Băile Tuşnad (Tusnádfürdő) of 26 July 2014. *The Budapest Beacon*. https://budapestbeacon.com/full-text-of-viktor-orbans-speech-at-baile-tusnad-tusnadfurdo-of-26-july-2014/.

Vári, A., 1997. A környezeti döntésekben való társadalmi részvétel és konfliktuskezelés fejlődése Magyarországon, in: Kárpáti, Z. (Ed.), *Társadalmi és Területi Folyamatok Az 1990-es Évek Magyarországán*. MTA Társadalmi Konfliktusok Kutató Központja, Budapest, pp. 273–97.

Vay, M., 2004. *Meddig vagyunk? — Válogatott írások a Védegyletről*. Noran Kiadó, Budapest.

Watts, M., Peet, R., 2004. Liberating political ecology, in: Peet, R., Watts, M. (Eds.), *Liberation Ecologies: Environment, Development, Social Movements*. Routledge, London and New York, pp. 3–47.

2. The Making of the Environmental and Climate Justice Movements in the Czech Republic

Arnošt Novák

Ever since the environmental movement formed and emerged in the late 1960s and early 1970s in the USA and western Europe, it has been a significant social actor on the national and world stages. It has never been homogenous and monolithic, but has always been a conglomerate of various approaches and trends, tactics and strategies. Often these could be, and indeed were, in conflict with one another. A number of authors see environmentalism as having come in three waves: as a conservation movement from the late nineteenth century, a political environmentalism from the late 1960s and early 1970s, and finally as a form of radical environmentalism that appeared in the early 1990s (Doherty, 2002; Carter, 2007; Saunders, 2012).

The second wave politicised many environmental issues and brought them into general public discourse. Civil society organisations such as Greenpeace and Friends of the Earth also started to use unconventional repertoires of action, such as non-violent and symbolic direct action, to draw attention to various issues such as industrial pollution, nuclear energy, economic growth, or the global dimension of environmental risks. But during the 1980s and at the beginning of the 1990s, the mainstream environmental movement had been de-radicalised and institutionalised (Macnaghten and Urry, 1998). Once 'alternative' seekers and critics of a system based on growth, environmentalists transitioned to become fixers of neoliberal systems with the worst excesses. Many dominant actors

 https://doi.org/10.11647/OBP.0244.02

within the environmental movement participated in the creation of a seemingly depoliticised environment, where the means of governance, based on a neoliberal consensus, foreclosed political negotiation and discussion through dissensus, different values, practices, and visions of other worlds. Erik Swyngedouw has termed this foreclosure of alternatives "post-politics", where "[p]ost-politics reduces the political terrain of the sphere of consensual governing and policy-making, centred on the technical, managerial and consensual administration (policing) of environmental, social, economic or other domain, [where] they remain of course fully within the realm of the possible, of existing social relations" (Swyngedouw, 2011: 266).

Critically distancing itself from this development, the third 'radical' wave of environmentalism and environmentalists emerged at the beginning of the 1990s (Carter, 2007; Saunders, 2012), when ever-greater numbers of activists reported that organisations such as Greenpeace and Friends of the Earth had limited the space for grassroots participation and decisions, and had become 'chequebook organisations' that asked people to contribute financially, on the understanding that organisations would then act on their behalf. This wave of radical environmental activism was characterised by a shared conviction and determination to confront and unmask those who were damaging the environment, as well as by an emphasis on participation in how environmental activities were designed and undertaken (Wall, 1999; Seel, Paterson, and Doherty, 2000).

Throughout heterogenous Western environmental movements, these changes in environmentalism can be seen as cycles of radicalisation and de-radicalisation. Since 2009 similar cycles have affected the climate movement, and there have been recent shifts towards radicalisation as evidenced by the concept of "climate justice", which aligns the fight against climate change with the fight for (local and global) justice issues (Cassegard et al., 2017).

By contrast, environmentalism and environmental activities in the Czech Republic have a different historical trajectory from environmental movements in western Europe and the United States. Eastern European (EE) movements have always been inspired by and oriented in relation to Western environmental movements. When the former Czechoslovakia was governed by authoritarian state socialism, environmentalism consisted essentially of only apolitical, state-sanctioned conservation organisations. These organisations tried, above all, to awaken an interest in the environment on an everyday level through educational activities and minor conservation works such as cleaning up streams and woods, collecting rubbish and scything meadows (Fagan, 2004). "Czech environmentalism in the late 1980s was a strange mixture of the officially-sanctioned moderate current of rational, technocratic and scientific thinking about environmental problems with an unofficial, romantic undercurrent that praised the beauty of untouched nature and individual freedom" (Jehlička et al., 2005: 75). This romantic undercurrent was cultivated mainly in leisure activities, such as tramping, which was a distinctive subculture of escaping into the woods and countryside. These lifestyle activities, tolerated by the regime, coexisted well with a conservation ecology that was based on rational, technocratic and scientific thinking. Both traditions shared a dualistic conception of nature and society and did not present a critique of modernity in the spirit of the Western New Left counterculture, thus preparing the ground for the onset of ecological modernisation in the 1990s (Jehlička and Smith, 2007).

The beginning of the 1990s witnessed in the Czech Republic the formation of two important environmental organisations: the Rainbow Movement and Children of the Earth. These organisations were inspired by the second, political wave of environmentalism led by Greenpeace and Friends of the Earth, which made use of a repertoire of protest actions such as non-violent direct actions and blockades, and criticised the accelerated transformation of society towards capitalism at the expense of the environment. In contrast to previous, rather apolitical forms of conservation, these organisations tried to focus on the roots of environmental problems; they criticised excessive consumerism, economic growth, free markets, and connected local issues with global ones. However, since the mid-1990s, a gradual de-radicalisation, institutionalisation and professionalisation of environmental organisations took place, leaving aside more radical tactics and cosmologies (Novák, 2020). Instead of proffering a fundamental critique of consumerism, growth, capitalism and a search for alternatives, moderate changes within the legal system and corrections to the neoliberal model were sought by the same actors who had searched—a decade previously—for alternative developmental pathways with less harmful environmental consequences. Thus, Czech environmental organisations transformed themselves from alternative seekers to system fixers. From the end of the 1990s until 2015, when the climate justice group *Limity jsme my* (We are limits) was founded, the origins of a third wave of domestic environmentalism took place, consisting of several small, professionalised ENGOs that operated on the basis of a transactional activism without any informal activities, confrontational repertoires or posed challenges to dominant powers, which are the supposed key characteristics of any social movement (cf. Doherty, 2002).

Why this process of de-radicalisation and institutionalisation over more than twenty years? What has caused the reversal of this process? Why can we now see not only the 'third wave' of radical environmentalism in the Czech Republic, but also the radicalisation of moderate ENGOs of the second wave? The following sections attempt to explore and find answers to these questions.

Methodology

The empirical basis for this chapter is based on in-depth formal and informal interviews, observations and analysis of documentary materials and written texts. Over the period 2010–2015, I carried out a total of twenty-five in-depth interviews with both present-day and former activists from Czech environmental organisations. I also drew on my past involvement in environmental protests: with this relatively extensive experience, I made it subject to analysis within the framework of memory work (Berg, 2008) and as a basis for further research, which enabled me to increase my theoretical sensibility towards research questions and interpretations. Although I have never been a member of any ecological or environmental organisation, since the early 1990s I have been associated with the counter-cultural anarchist scene that occasionally overlapped with ecological activities, especially during the blockades of the demolition of the village of Libkovice due to coal mining, and in protests against the completion of the Temelín nuclear power plant. I thus took part in a range of environmental protest actions from the beginning of the 1990s. Since 2014, my research has been based primarily on ethnographic methods and participatory observations of meetings and protests, as well as informal interviews.

My experience as an activist allowed me to take a look at my research topic from an angle that is different from the more common approaches found in the Czech academic field. Moreover, it permitted me to ask questions that would not occur to an academic lacking an embedded, grounded experience. It also positioned me well to comprehend the context of information obtained from respondents and to analyse materials throughout the research. This work thus documents my position at the interface between two distinct worlds: first, the academic, as one with its inclination towards an objectification of knowledge and its requirement for researcher detachment from the object under investigation as well as from his values; second, the activist, where all actions are encumbered with values and emotions, who is much less concerned with describing the universe than with changing the world. I drew inspiration from British social geographer and Earth First! activist Paul Routledge who came up with the 'third space' approach that "implies inappropriate(d) encounters between academia and activism where neither side, role or representation hold sway, where one continually subverts the meaning of the other" (Routledge, 1995: 400).

The third space as critical engagement has allowed my role as an academic and my role as an activist to constantly disturb one another constructively, with one critiquing the other, cultivating criticism and greater or different understandings of the explored topic, and in a way rectifying the 'unfeeling' detachment of an academic on the one hand and the 'passionate' preoccupation of an activist on the other.

Czech Radicalisation from the Early 1990s

The beginning of the 1990s witnessed, in the Czech Republic, the formation of two important environmental organisations — *Hnutí Duha* (Rainbow Movement) and *Děti Země* (Children of the Earth). Traditional conservationists during late state socialism ascribed the cause of environmental devastation to central planning systems and directive state government, and looked with hope towards parliamentary democracy and the market. In contrast, new organisations were much more critical of the latter and open to alternatives. *Děti Země* did not aim to form a common worldview—instead, the group was an open platform for meeting and seeking alternatives to societies actively destroying the environment and using up finite resources. People of different kinds gathered under its flag—from ecologists and counter-cultural people (mostly punk), to anarchists. On a political level the group was somehow moderated: it focused on local contested issues around, for example, transportation or extractivist companies without engaging in explicit systemic critique. The group was moderate but at least at the beginning of the 1990s, it occasionally made use of blockades and demonstrations:

> We didn't want to change a system, but focused more on influencing politics of the environment. We wanted to change a system *attitude* but we did not care much what kind of system it was. Obviously, a dictatorial regime does not like even these politics, so there were limitations (an interview with one of the founders of *Děti Země*, 2011)

Děti Země represented a classical environmentalism of the second wave, which contained, as in the West in the 1970s and 1980s, rudiments of a radical environmentalism, wherein systemic alternatives are sought and proposed, for example as critiques of consumerism, industrialism or humans' domination of nature. Above all, the group wanted to play the role of 'watch dog'. This is where it differed from *Hnutí Duha*. Jakub Patočka, one of its founders, represented a charismatic authority with a certain vision that he tried to enforce through the movement. Because of him, *Hnutí Duha* brought a political dimension into the Czech conservation-dominated environmentalism. Arguments over ecological topics started to be politicised in the public and media sphere, instead of being grounded only in scientific and technocratic expertise (Lee, 1995).

For example, *Hnuti Duha*'s journal *The Last Generation* published pieces about new topics and sometimes radical opinions: thematised repeatedly were the activities and the negative impact of huge financial institutions and trans-national corporations on local communities, nature and the environment. The journal also transmitted news and information from Western environmental magazines such as *Third World Resurgence*, *The Ecologist* and *The Resurgence*. *The Last Generation* therefore constituted an important source of new topics for discussion within the Czech Republic, where economic growth, the rising power of global corporations, and consumer society were criticised along with party-political systems.

The Rainbow Movement introduced new repertoires of actions such as 'happenings', direct actions, blockades and demonstrations. Mainly because of these activities, the Rainbow Movement was labelled by the media as a 'radical organisation'. Inspired more by Greenpeace than radical ecologists from Earth First!, the Rainbow Movement and other Czech environmental activists operated with a liberal notion of direct action (Duckett, 2006). Duckett characterises liberal direct action with three features: first, it is understood to be the ultimate possibility at hand when all other 'accessible' options have failed. Second, it is an efficient form of lobbying and tactics through which one can bring media and public attention to issues, and thus create pressure and involve new actors in decision-making. Third, liberal direct action needs external, 'democratic' legitimation and should therefore be used when state offices or other institutions break laws or act irresponsibly (Duckett, 2006: 155). In addition, liberal direct action is strictly non-violent and to a certain extent symbolic, because "...direct action is not a solution, neither a goal on its own, but it is a unique instrument in how to bring attention to topics" (Monbiot, 1998: 185).

It was precisely this 'liberal' concept that was implemented in the Czech environmental movement and systematically developed in trainings, movement materials, everyday praxis and in the biggest environmental campaigns of the 1990s, where the identity and public image of the environmental movement was formed. Here, I refer specifically to campaigns such as those to save Libkovice village from destruction owing to coal mining at the turn of 1992 and 1993, or those against the end-phase of the construction of the Temelín nuclear power-plant between 1995 and 1997. In both campaigns, and especially in the case of Temelín, many different worldviews such as ecological, subcultural (mostly punk), and anarchist merged as people met during blockades. However during these actions, the instrumental use of politics in the form of the liberal concept of direct action clashed with the anarchistic concept of direct action, which has also been one of the characteristics of third wave radical environmentalism. The blockades at the construction site of Temelín were organised by the leadership of

Hnutí Duha, which used direct action as a form of lobbying, and as a way to bring media and public attention to the issue. However, the blockade was organised hierarchically, and people were left feeling that they were an "infantry under command" (an interview with one of the former activists of *Hnutí Duha* and a later activist of *Nesehnutí*, 2011), without possibilities to participate in decision-making. Moreover, the leadership of *Hnutí Duha* wanted not only to dominate the blockades but also to be hegemon of the anti-nuclear movement. This instrumentalisation of direct actions, and generally the battle between environmental groups for dominance and power, created tensions with more anarchistic segments of the anti-nuclear blockades.

With regard to the conservation-focused environmentalism from before 1989, *Hnutí Duha* and to a lesser extent *Děti Země* became Czech versions of 'radical environmentalisms'. In this context, these groups used a non-conventional repertoire of actions such as blockades anchored in the liberal notion of direct action (no matter how conventional these actions were in the West as part of a moderated second wave), because under the previous authoritarian state socialist regime, this kind of action—and even common street demonstrations—was banned.

This environmentalism brought to a rapidly transforming post-socialist society a strong critique of consumerist society, of trans-national corporations, and economic growth, as well as questions and new ideas about the valuation of nature itself. Precisely in this context the activities of these groups, and especially of *Hnutí Duha* in the first half of the 1990s, could be understood as radical. In a society under transition and intent on 'catching up' with the West, these environmental actors sought to carefully seek alternatives—not openly and directly to capitalism, but rather to consumer society. 'Catching up' in this time applied not only to the economy and governance, but to environmental thought, critique and practice.

De-Radicalisation

As mentioned above, during big ecological campaigns such as those at Libkovice or Temelín, many different kinds of people merged, including anarchists and punk rockers. Even though it was common at the beginning of the 1990s for environmentalists and anarchists to co-operate, from the end of 1993 onwards, Jakub Patočka, the charismatic founder of *Hnutí Duha*, started to strongly disagree with coalitions formed with anarchists because he disagreed with their ideas, alternative cultures (he considered punk to be an electronic noise) and their "unclear relation to violence" (Patočka, 1993). There was also a very practical reason for this refusal to cooperate: anarchists were much stronger in numbers than *Hnutí Duha*, and they could conceivably have overtaken the movement as an ecological flagship for their own anarchism and taken it in an anarchist direction, thus changing the entire remit and objective of environmentalism. This threat was likely the real reason to "recall certain distance and differences" (Patočka, 1999: 20): from the mid-1990s, Patočka started to build strong boundaries between the Rainbow Movement (and thus the entire environmental movement, in which the *Hnutí Duha* was about to play a hegemonic role), and the anarchist movement. In the 1990s, anarchists were one of the few opponents of skinhead youths' racist violence and were willing to confront them violently in the streets. It was mainly due to these clashes, portrayed by the media as clashes between two gangs, that they were called violent. In addition, together with their criticism of the state and capitalism, they were labelled as left-wing extremists by police and the media, but also by some of the environmental NGOs.

> Jakub Patočka defined on behalf of *Hnutí Duha* and the ecological movement in general why it was unacceptable to cooperate with anarchists — the main reason was that anarchists had an unclear relationship to violence, whereas the ecological movement was strictly non-violent. There were also other kinds of arguments apart from the violence question, but the friction was seen to focus mainly around practical issues. When anarchists took part in *Hnutí* actions, no matter if they acted violently or not — and I do not remember where they were violent — they were put on notice that their participation was not welcomed (interview with activists from *Hnutí Duha*, 2011).

Another tension was produced by the internal hierarchy and non-democratic structure of the movement, combined with its authoritarian leadership. This became evident in the hierarchically organised blockades of Temelín. Patočka's rejection of anarchists led to a reaction from the anarchists, who started to distance themselves in turn from *Hnutí Duha*, and the whole NGO sector. In the mid-1990s, a dogmatic version of social anarchism became stronger and stronger and contributed to firm boundaries between environmentalists and anarchists. Groups such as the Federation of Social Anarchists started to promote an opinion that refused any cooperation with environmental and non-governmental organisations and labelled them, together with anarchists willing to cooperate, "collaborators": "The task, [...] is foremost to untie every relation with democratic, 'green' collaborators, because it is precisely these organisations that bring anarchists to the edge of a cliff" (PW, 1998: 3).

Structural factors strengthened these boundaries as well. After the victory of Václav Klaus in the parliamentary election of 1992, environmental organisations started to be perceived as a brake that was slowing down, or preventing, the country from catching up with the West. The winners of the Velvet Revolution were gradually grouped on the side of losers: this process culminated in 1995, when these groups were put on the Ministry of Interior's list of extremist organisations (Fagan, 2004).

It was important for non-governmental environmental organisations, to start to prove their non-extremist positions by stressing their modesty, as their reputations were fragile and problematic considering that they had previously—albeit briefly—partnered and linked up with, through donors and media image orientations, anarchists officially labelled as extremists. Environmentalists actively distanced themselves by emphasising their non-violence and their constructive approach from within the system. Greenpeace Czech Republic decided for example to follow a strategy that emphasised scientific argumentation and expert knowledge at the expense of emotional presentations (Jehlička, 2009) from this time on:

> *Hnutí Duha* started in those times to advance lobbying methods as well as a strategy of cooperation with some parties. The movement published recommendations around which party to vote for and which to avoid. During lobbying for laws, personal ties emerged. And the common argument was that when we cooperate with anarchists, there is no chance to advance lobbying (interview with an activist from *Hnutí Duha*, 2011).

Distancing from the anarchists and a refusal to cooperate with them became part of the self-disciplining that environmental organisations undertook during the consolidation of liberal capitalism in the country. Even though placing such groups on the extremist list actually helped to normalise their existence, as they did not perceive themselves (nor act) as challengers of the status quo, it did not prevent a discourse of environmentalism as a threat to freedoms and economic development from emerging (Klaus, 2007; 2009).

One can trace the shift within the *Hnutí Duha* agenda from more radical alternative seekers to 'system fixers' to between the years 1997 and 1998. In 1997, the movement stopped organising blockades of Temelín and direct actions generally. Their self-presentation in annual reports changed as well. Between 1995 and 1997 one could read that industrial civilisation was unsustainable, that economic growth was an undertaking to be criticised together with a particular vision of 'progress', and that non-violent direct actions were legitimate. The goal of *Hnutí Duha* at this time was not to give rise to shallow, inconsequential changes, but to highlight and withdraw from the roots of systemic problems, to challenge and change societal values away from consumption to frugality, and to enable societal change through the promotion of localised, deliberative decision-making processes closer to people affected by interventions and decisions, wherein local communities should be empowered and the importance of economic aspects mitigated by other relevant considerations.

The change in reports come from 1998—the first year that *Hnutí Duha* did not organise or take part in the Temelín blockades, and instead adopted a lobbying strategy. The first sentence from the report modified and made clear the new aims of the organisation, stating that its goal was to advance nature's recovery, to change people's values from narrow-minded consumerism towards a democracy of engaged citizens. *Hnutí Duha*'s efforts came to be concentrated around everybody living well on Earth. Another year later, aims to change values of people and society were replaced with aims to make efficient and realistic acquisitions that would enable limitations on and decreases to pollution. One can still read about system arrangements, but no longer in the context of unsustainable society and civilisation trends, and rather in the context of reforms of mineral mining charges, reforms required to a packaging law, or as part of the liberalisation of the electricity market (Binka, 2008: 136–37). Thus, the Czech environmental 'movement' through *Hnutí Duha* as hegemon definitively moved away from a politically-supported basis of negotiation, discussion and dissensus to consensual post-politics within a post-socialist version of neoliberalism. This trajectory was similar to their inspirational patterns of second wave environmentalism, but it lasted for a shorter time.

The Czech environmental movement thus went through a similar cycle to western movements in the '90s, albeit with a delay: from radicalisation to introducing nature into politics, to institutionalisation, de-radicalisation and co-optation of the movement by the capitalist status quo. The difference was that no 'third wave' radical environmentalism subsequently replaced it, and instead of a multi-dimensional social movement, several mainstream, professional advocacy ENGOs focused on lobbying and negotiating with the state administration and the commercial sphere became dominant after 2000. This consensual depoliticisation was enhanced in 2004 when the Czech Republic joined the EU, as the EU urged a particularly technical approach to the environmental agenda. Thus, laws to protect nature were not an outcome of compromise as a result of mobilisation and pressure from below, but from above, predominantly from Brussels by means of bureaucratic directives.

'New Wave' Radicalisation Post-2015

This state of affairs changed around 2015, when the *Limity jsme my* (LJM) movement was formed. This began with campaigns and protests against coal mining, and in 2017 the LJM organised the first climate camp through mass direct action to occupy a coal mine. LJM was predominantly formed as a result of activists' experiences in German climate camps organised by Ende Gelände. They represent an example of radical climate movement activism, by which I understand the struggles for "System Change, not Climate Change" as referring to the need to overhaul the political and economic system that fundamentally gives rise to climate change (Temper, 2019).

The climate camps and activities of LJM not only emphasised non-hierarchical horizontality and a repertoire of activities that focused on mass direct actions, but also the adoption of a system-level critique that tried to politicise environmental issues and the climate crisis by aiming

Fig. 1. Direct action of LJM in 2017 (occupying the mine). Photograph by Majda Slámová (2017).

at the systemic nature of the effects of climate change and the ways in which it was connected with political power and the fossil fuel industry. The movement began to reunite people from different backgrounds, including anarchist and counter-cultural ones as in the first half of the 1990s. In contrast to the previous era, much more effort was put into capacity-building to create a non-hierarchical grassroots movement. An important role in the intermingling of different milieus was played by Klinika, the autonomous social centre in Prague, which at the time of the formation of *Limit jsme* served not only as an inspiration for radical activism, but as an important space for encountering these different milieus (Novák and Kuřík, 2020; Novák, 2021).

In their manifesto they wrote:

> We are part of a global movement [...] We feel solidarity with people mostly affected by climate change and ecological problems. We are aware that ecological problems cannot be separated from social and economic issues [...] We work non-hierarchically, we do not have bosses, we are independent of political parties. [...] We stand for direct action. Radical politics, which intends to solve deeper causes of social problems, has never been successful without direct action [...] Direct action is our main,

> although not exclusive, means of changing and mobilising society [...] We want to change the system, not the climate!" (Manifesto LJM, 2017)

To a much greater extent, the new environmentalism is aligned with the anarchist concept of direct action, however, it is not understood instrumentally. An anarchist direct action is not primarily instrumental, but a goal and a value in itself:

> Direct action is not only a tactic; it is an effort of people to enforce control over one's life and to participate in social life without the need of mediation or control from the side of bureaucrats and professional politicians. Direct action places a moral commitment over a positive law. Direct action is not the ultimate possibility, when all other possibilities fail, but it is a preferred way of action (Doherty, Plows, Wall, 2003: 670).

Unlike the 1990s, the contemporary movement's organisers are not burdened with the dogmatic reasoning of a 'pure' radical politics and are able to think much more strategically. For this reason, they are able to cooperate with mainstream ENGOs even if they disagree on broader issues or tactics. When the government set up a coal commission (to talk about the end to and transition away from coal), where the commission's composition favoured the fossil industry, ENGO representatives decided to accept the invitation. LJM criticised this decision, but at the same time accepted their own invitation as they reasoned that it was a route to putting even more pressure on the commission from below (which is a common strategy with moderate ENGOs).

Through capacity-building (often with the cooperation of moderate second-wave organisations such as Czech Greenpeace or *Hnutí Duha*, as they use their infrastructures), LJM have also mobilised more people, with a growing number participating at the 2018 and 2019 climate camps. In March 2019 Czech high school students took part, for the first time, in a large number of strikes that helped bring the issue of climate change into the public sphere. Several founders of the Czech 'Friday for the Future' strike had previous experiences with LJM's climate camps.

The year 2019 was important because, as in Great Britain, the Extinction Rebellion (XR) movement originated and organised several civil disobedience actions. Unlike LJM, which focuses on blockades of fossil fuel infrastructure and system-based critiques of economic growth or occasionally fossil capitalism, XR have chosen a different strategy. Specifically, they have focused on blockades around individual transport aimed at disrupting people's regular daily routines and thus forcing them to take an interest in the climate crisis, particularly as XR target the effects and influence of fossil corporations and their associated infrastructure (mines, powerplants). XR have also chosen a different communication line. Their moralising arguments mention consumerism more and focus on rousing and motivating individuals into action and participation. This corresponds with their tactics of blockading car and bus traffic in Prague. This differing strategy has stirred up frictions within the climate activism movement in the Czech Republic and provoked disputes about tactics and strategy. In contrast to the '90s, this has not led to exclusionary disagreement and differing trajectories between groups.

Thus, in 2019 a climate movement that is more than just environmental has emerged. It is not only centred around a few NGOs that function according to the principle of transactional activism. Besides climate camps and movements trying to articulate a radical system critique and re-figure society for climatic equality, new organisations are emerging. Secondary school strikes such as Fridays for Future (FFF), which appeal to and target the government and key decision-makers on the issue of climate change, or the Czech branch of XR, combine a moralising tone with an occasional apocalyptic undertone. Together with NGOs like *Hnutí Duha* or Greenpeace, these are beginning to constitute a movement that is diverse in terms of climate change causes, analysis and potential response strategy. LJM are striving to break post-political conditions and to re-politicise the topic of climate change and instigate mobilisation from below: their grassroots political activism mobilises predominantly younger people without previous activist experience. XR, through their deliberate depoliticisation and lack of engagement with 'left' or 'right' politics, and their adoption of moralistic and apocalyptic argumentation, are able to mobilise people of diverse ages as well as those without any prior experience of activism.

All environmental groups share an inspiration by and continuation of the practice of environmentalism and protest tactics from abroad. LJM were inspired by the German Ende Gelände as they try to re-shape the inspiration to fit the Czech context. They use tactics such as mass direct actions in white overalls and adopt a strategy that focuses on fossil infrastructures, combined with systemic criticism, and as a result of its post-socialist context, without strong or explicit critique of capitalism. Secondary school students are picking up the threads and example of Greta Thunberg and FFF abroad, as are Czech XR from their UK founders. Some XR participants label themselves as a branch of the English founders; others find re-shaping XR's concept from the English to the Czech context quite schematic and unproblematic. Nevertheless, this orientation to foreign environmentalism helps us realise the extent to which these differing strands of the present climate movement relate to politics. While LJM represents activism that is not state-oriented (as they are beyond the state) as they try to formulate alternative political visions by their actions (degrowth, democratisation of power engineering and climate justice), FFF and XR are much more state-oriented as they target, address, and invite governments to act. This approach arguably threatens to keep solutions to the climate crisis within the boundaries of the present capitalist status quo. As FFF also argue that we should abide by the assertions of scientists, they contribute to the further depoliticisation of the climate crisis as they reify and re-frame it as a technical problem. Like XR, the depoliticisation of knowledge-making systems and proffered scientific 'solutions' displaces alternative visions of social organisation as they emphasise and elicit fear, apocalypticism and moralising around individuals' behaviour.

In any case, 2019 has brought remarkable activity and change to within the environmental and climate movement, which is becoming a major social player once again. These developments have given rise to a space in which debates and conflicts around environmental futures may be newly, differently, and deliberatively negotiated. One might say that in 2019, the environmental movement and the climate movement have become a social movement (Doherty, 2002). It has a consciously collective identity, with activists acting partly outside political institutions, using protest as one of its forms of action, and is characterised by un-institutionalised networks of interaction as mainstream institutions are rejected by large parts of the movement so as to better challenge dominant forms of power.

It is currently a movement with great momentum and potential. Despite their confrontational repertoire of action (which has not previously been very popular within Czech society) such as the

Fig. 2. Blockade of main roads in the centre of Prague, action of XR in 2019. Photograph by Petr Zewlakk Vrabec (2019).

occupations of mines, strikes and blockades, climate groups have garnered public support. According to a survey from December 2019, sympathy for high school climate strikes in the Czech Republic slightly outweighed those who disliked them (36% vs. 33%); sympathy for the *Limity jsme my* initiative was significantly more positive than negative (37% vs. 28%); and only in the case of Czech Extinction Rebellion were respondents significantly more unsympathetic (49%) than sympathetic (28%) (MEDIAN, 2019).

Conclusion

Based on these insights and review, it is possible to say that the composition of environmental activism has changed from a narrow, topic-based, and rather apolitical conservation approach during the era of state-socialism, to a brief, politically-inclusive upswing that examined the possibilities of a consumerist-free market society at the beginning of the 1990s, to its deradicalisation towards the end of the '90s when environmentalists became 'system fixers' of the post-socialist version of capitalism. Subsequent activism around climate change re-opened what could be debated, negotiated and done, and has today brought about radical critique. Debate about systemic change is conducted increasingly in the public sphere together with a non-conventional repertoire of civil disobedience and direct action, as both these approaches embrace a more participative activism.

There are several possible reasons to explain why this long-term process of de-radicalisation, institutionalisation and de-politicisation has been reversed by a cycle of radicalisation and politicisation. There are some external structural reasons, including a question of agency and also the interplay between these factors. Climate change is increasingly understood not only as an ecological, but also as a social, problem. The contemporary third wave of environmentalism in the Czech Republic is more akin to a social-ecological movement that refuses and disputes depoliticised environments that originate from technical, managerial and consensual administrations and experts (cf. Swyngedouw, 2011). Second, after the failure of the Copenhagen COP in 2009, some moderate organisations radicalised and shifted their focus towards the concept of climate justice. The Czech Greenpeace and *Hnutí Duha* (as part of Friends of the Earth International) were part of this international debate and shift. Third, neoliberal hegemony collapsed in the country after the financial crisis of 2008–2009. The protests against the austerity politics of the right-wing government in 2011 and 2012 in the Czech Republic were a manifestation of this collapse. Subsequently, a new generation of young people, especially at universities, has been brought to politics through these protests. After the 2013 fall of the government, many became politicised and continued their protest activities within the environmental and climate justice movement.

These factors and events together mean that radicalisation and politicisation of the movement is not only due to political opportunities and windows, but is also a result of the agency of individuals and groups and their perceptions and strategies to communicate the existential threat and urgency of modern environmental crises and challenges. These have culminated in the intentional creation of a different kind and way of doing politics.

References

Berg, A. J., 2008. Silence And Articulation — Whiteness, Racialisation And Feminist Memory Work, *NORA — Nordic Journal of Feminist And Gender Research*, 16 (4), 213–227. https://doi.org/10.1080/08038740802446492.

Binka, B., 2008. *Zelený extremismus*. MUNI PRESS, Brno.

Carter, N., 2007. *The Politics of the Environment. Ideas, Activism and Policy*. Cambridge University Press, Cambridge.

Cassegard, C., et al., 2017. *Climate Action in a Globalizing World*. Routledge, London.

Císař, O., 2009. Social Movements and Political Mobilization in the Czech Republic after 1989. *Transition*, 49 (1), 33–46.

——, 2008. *Politický aktivismus v České republice. Sociální hnutí a občanská společnost v období transformace a evropeizace*. CDK, Brno.

Doherty, B., 2002. *Ideas and Actions in the Green Movement*. Routledge, London.

Doherty, B., Plows, A., Wall, D., 2003. The Preferred Way of Doing Things: The British Direct Action Movement. *Parliamentary Affairs*, 56(4), 669–686. https://doi.org/10.1093/pa/gsg109.

Duckett, M., 2006. Ecological Direct Action and the Nature of Anarchism: Explorations from 1992 to 2005, Newcastle University (Unpublished doctoral thesis).

Fagan, A., 2004. *Environment and Democracy in the Czech Republic: The Environmental Movement in the Transition*. Edward Elgar Pub, London.

Jehlička, P., 2009. Recenze: Ondřej Císař: Politický aktivismus v České republice: Sociální hnutí a občanská společnost v období transformace a evropeizace. *Mezinárodní vztahy*, 44(4), 112–117.

Jehlička, P., Sarre, P., Podoba, J., 2005. The Czech Environmental Movement's Knowledge Interest in the 1990s: Compatability of Western Influences with pre-1989 Perspectives, *Environmental Politics*, 14(1), 64–82. https://doi.org/10.1080/0964401042000310187.

Jehlicka, P., Smith, J., 2007. Out of the woods and into the lab: exploring the strange marriage of American woodcraft and Soviet ecology in Czech environmentalism, *Environment and History*, 13(2), 187–210. https://doi.org/10.3197/096734007780473546.

Macnaghten, P., Urry, J., 1998. *Contested Natures*. SAGE Publications, London.

MEDIAN, 2019. Attitudes of the Czech public towards the energy concept and coal mining, MEDIAN, December 2019, https://www.median.eu/en/wp-content/uploads/2020/04/MEDIAN_energeticka_koncepce_tezba_uhli_200210.pdf.

Monbiot, George, 1998. Reclaim the fields and country lanes! The Land is Our campaign, in: McKay, G., (ed.), *DiY Culture. Party and Protest in Nineties Britain*. Verso, London.

Novák, A., Kuřík, B., 2020. Rethinking radical activism: Heterogeneity and dynamics of political squatting in Prague after 1989, *Journal of Urban Affairs*, 42(2), 203–21. https://doi.org/10.1080/07352166.2019.1565820.

Novák, A., 2021. Direct Actions in the Czech Environmental Movement, *Communist and Post-Communist Studies*, 53(3), 137–56. https://doi.org/10.1525/cpcs.2020.53.3.137.

——, 2021. Every city needs a Klinika: The struggle for autonomy in the post-political city, *Social Movement Studies*, 20(3), 276–91 https://doi.org/10.1080/14742837.2020.1770070.

Patočka, J., 1999. Duha deset let na cestě: lidé a křižovatky, *Sedmá generace*, 8(10), 15−23.

——, 1993. Anarchisté a Hnutí Duha, *Autonomie*, 16: 12.

PW, 1998. Anarchismus — nutně sociální, nutně revoluční, nutně kolektivistický a racionalistický, *Svobodná práce*, 12, 3.

Routledge, P., 1996. The Third Space as Critical Engagement, *Antipode*, 28(4), 399−419. https://doi.org/10.1111/j.1467-8330.1996.tb00533.x.

Temper, L., 2019. Radical Climate Politics. From Ogoniland to Ende Gelande, in: Kinna, R., Gordnon, U. (Eds.), *Routledge Handbook of Radical Politics*. Routledge, London.

Seel, B., Paterson, M., Doherty, B., 2000. *Direct Action in British Environmentalism*. Routledge, London.

Swyngedouw, E., 2011. Depoliticized environments: The end of nature, climate change and the post-political condition, *Royal Institute of Philosophy Supplements*, 69, 253–74. https://doi.org/10.1017/s1358246111000300.

Wall, D., 1999 *Earth First! and the Anti-Roads Movement*. Routledge, London.

3. The Construction of Climate Justice Imaginaries through Resistance in the Czech Republic and Poland

Mikulás Černìk

Today, the use of coal—and control over it as a resource—is increasingly contested. Environmentalists tend to demand a post-carbon transformation, while the coal industry and the state seek to secure the continued use of coal. According to Huber and McCarthy (2017), subterranean fossil resources enabled the shift from a derivation of power through control over land and territory, to its derivation through control over machines and labour. Yet regions of fossil fuel extraction remain crucial for the power of the state, where these regions carry at the same time the unevenly distributed negative consequences of mining (Frantál and Nováková, 2014). State energy sovereignty based around domestic production of coal fosters local extractivism, resulting not only in the destruction of the environment but also in a comparably worse quality of life in the regions of extraction.

Resistance to coal mining can alleviate or even reverse this extractive relationship. Resistance may lead to an articulation of views and alternatives around futures without coal, not only locally in coal-producing regions, but also by highlighting and questioning the uneven impacts of climate change caused by coal combustion. The nexus between energy metabolism and political and economic power has a 'glo-cal' nature, exemplified by Klein's term "Blockadia" (2015), which connects local cases of resistance to fossil infrastructure as part of the same movement. Local conflicts are not just isolated cases but have global, political, and economic implications. They show the

 https://doi.org/10.11647/OBP.0244.03

power of participants to intervene in the mitigation of climate change, an endeavour in which global leaders have failed. Yet, in each place, resistance creates specific risks to local power structures, giving voice to new groups and creating new dynamics and dependencies—and also provokes new state responses and regulatory powers.

In this chapter I am following the narrow definition of an overt resistance mentioned by Hollander and Einwohner (2004). Resistance is not only an intended act—it also has to be recognised as resistance by observers. Protests against coal mining—as acts of resistance—politicise nature-society relationships, and can be seen as acts of citizenship (Rasch and Köhne, 2016), but also provide a distinction between civil and uncivil actors (D'Alisa et al., 2013). As with other ecological distribution conflicts, resistance to coal mining could become a driver of sustainability, but also a driver of climate change mitigation (Temper et al., 2018). "Politics occurs wherever a community with a capacity to argue and to make metaphors is likely" (Rancière, 1999: 60): the capacity to argue goes beyond mere participation in a public debate, such as setting demands for politicians, or calling for certain legislative procedures. It is the ability to create an argument as a new frontline of political conflict that cannot be ignored or disregarded. In the case explored here, it is an expression of disagreement and discontent towards the corporate-state-mining complex which fosters the dependency of energy metabolism on coal. Such 'arguments' are a vital part of the demands of a radical democracy with alternative imaginaries (Lloyd, 2009).

This chapter provides an overview of the protest events organised by climate justice movements across two regions of lignite extraction in Poland and the Czech Republic—namely the North Bohemian Coal Basin and the Konin Basin. Although we can talk of a global climate justice movement, its manifestation, actions and power to affect energy metabolism vary in different areas, and I will explore these in an eastern European (EE) context. We can understand the climate justice movement as a polycentric struggle, where simultaneous and mutually coordinated tactics and approaches are deployed around the globe and against various levels of governance (Tormos-Aponte and García-López, 2018). However, we can observe two different streams of the climate justice movement with regard to the hallmark of global climate politics, the United Nations Framework Convention on Climate Change (UNFCCC) summits. First, there are the reformist organisations, who observe and participate in the climate summits, known as Conferences of Parties (COP). The second, more revolutionary stream is intent on creating a different, radical and bottom-up politics. Climate camps are an example of the latter stream, which does not mean that they create a homogenous space. On the contrary, other authors observe divergent streams within the climate justice movement, such as leftists and environmentalists, anarchists and formal NGO participants, and all these groups have been present at the camps under consideration here (Reitan and Gibson, 2012; Saunders and Price, 2009; see also Novák, Chapter 2 in this volume).

Belonging to the broad and diverse tradition of protest camps (Frenzel et al., 2014), climate camps emerged in the UK in 2006 (Bergman, 2014; Russell et al., 2017; Saunders and Price, 2009), and during the last few years have been organised by climate justice movements in other European countries, notably in Germany. In EE, a climate justice movement has emerged only very recently. This is due to the specific situation of environmental activism in these countries, historically driven by the dependency on coal, on the one hand, and a prevalence of liberal environmentalism and transactional activism dependent on institutional funding on the other (Císař, 2010; see also Novák, Chapter 2). Actions of mass civil disobedience were relatively rare for environmental movements in European post-socialist countries. This chapter explores how the organisation and arrival of climate camps have changed this.

Below, I provide an overview of disruptive and constructive aspects of resistance during climate camps organised in the Czech Republic and Poland. To do so, I will first briefly describe the importance of coal for energy production in these countries. I will elaborate on the regional characteristics of extraction and energy policy in both states, as well as previous public mobilisations against coal. The empirical part is then divided into two sections. In the first, I will provide an overview of disruptive elements of the resistance, conducted against the power that controls the use of lignite reserves. In the second, I will focus on the aspects and conditions through and in which the movement creates a space for new, counter-hegemonic imaginaries.

The Importance of Lignite for Poland and Czech Republic

The Czech Republic, Poland and Germany are called the "European coal heartland" (Osička et al., 2020) because of their importance to overall European coal production. Whereas in Germany state policies and contingent roadmaps have deliberately transitioned to phase out coal, both Poland and the Czech Republic are reluctant to follow this path. In these latter two countries a steady decline has been observed in recent decades in the share of coal within overall energy production, as well as employment. Since the late 1980s, the share of coal had decreased to 47% of installed capacity by 2017 for the Czech Republic, and 70% of installed capacity in 2018 for Poland. The share of lignite is more significant in the Czech Republic than in Poland, where hard coal is a dominant resource (ERÚ, 2018; Macuk et al., 2019). In absolute numbers, coal production peaked in 1984 at around 100 million tonnes of coal a year in the Czech Republic and at 201 million tonnes a year in 1971 in Poland, which made it the biggest coal producer in Europe at that time, excluding the Soviet Union (Andersson, 1999; Kuskova et al., 2008).

State energy policies and the national long-term strategies of both countries are showing no clear plan to phase out coal. However, in Czechia, State Energy Policy (SEP) is factoring in a decline of 11–21% share of coal in energy production (MPO, 2014). Meanwhile, in Poland, SEP aims to maintain 60% of electricity generation from coal (Ministry of Energy, 2018). A recent study conducted by Climate Action Network Europe (Flisowska and Moore, 2019) shows a contradictory approach taken by both countries, whereby they are involved in funding for a 'just transition' while at the same time lack a clear plan for coal phase-out.

Resistance to lignite mining emerged in the interplay between opportunities to influence and eventually stop mining, and was exacerbated by the local effects and scale of physical mining activities. This situation is different across coal regions, not only because of the extent of coal reserves, but also because of the prospects of the lignite mining companies. Some mines and powerplants are more vulnerable to closure than others—this vulnerability is determined partly as a result of legislative procedures, partly due to their economic prospects, and to the role of public opposition. Therefore, it is important to describe the conditions in which climate camps are situated in the respective lignite extraction regions under consideration here.

Approximately 75% of overall lignite production in the Czech Republic comes from the North Bohemian Basin region, and most of the potentially extractable reserves of lignite are located there. Given the importance of lignite for domestic energy production, this makes the region crucial for the Czech Republic. There are three open cast mines operating in the area, two of which belong to a private conglomerate, Severní Energetická, which operates the ČSA mine, and its daughter company, Vršanská Uhelná, which operates the Vršany mine. The third mine in the region, Bílina, is operated by a national energy company, Severočeské doly (a daughter company of ČEZ). A renegotiated governmental resolution expanding the territorial limits to mining at Bílina mine will prolong mining there beyond 2035 and the overall amount of extracted coal is expected to be up to an extra 150 million tonnes of lignite ("Severočeské doly a.s.," n.d.).

In Poland on the other hand, the Konin Basin used to produce approximately 13 million tonnes of lignite a year, out of 53 million tonnes of lignite total (Konin Basin is one of four areas in total, out of which Belchatów Basin, with 41 million tonnes of lignite a year, is the biggest) (Mazurek and Tymiński, 2019). Lignite is not as important a fuel in overall installed capacity as it is in the Czech Republic. It constitutes a share of 21%, whereas hard coal accounts for 49% ("Energy Transition in Poland — Forum Energii," n.d.). Several open cast lignite mines are located in the Konin Basin, which are nearing the end of their reserves. The mines in operation at the time of writing are Jozwin, Drzewce and Tomislawice. Whereas Jozwin and Drzewce are close to full depletion of their reserves, the Tomislawice open-pit mine—the only open cast mine to open in Poland after it joined the EU—is facing legal obstruction. Altogether, lignite production in the region decreased to approximately 6 million tonnes a year and thus there has been pressure to open the Osczislowo lignite mine in the region. All of these belong to a private venture, ZE-PAK (Ministerstwo Energii, 2018). In sum, lignite mining is close to reaching its limits at several mines and there is a drive to expand the limits at existing open pits (at Bílina mine, for example) or to open new open-cast mines (for example at Osczislowo).

Although the importance of the lignite mined in both regions is very different for the overall energy metabolism of the respective states (in the North Bohemian Basin it is rather crucial, unlike in the Konin Basin), there are commonalities in their "coal culture". This term refers to the way that the region is strongly connected to the heritage of industrial lignite mining (Brown and Spiegel, 2019; Kuchler and Bridge, 2018). The term "corporate-state-mining complex" attends to the interconnectedness of market mechanisms and state power in public-private ventures, whereby the population is socially engineered through the creation of financial dependency on sponsorship and funding by mining companies on the part of municipalities (Brock and Dunlap, 2018). For example, fees and taxes distributed to the communities affected by mining provide an important source of finance, but they also play an important role in facilitating the approval of mining activity by local authorities and inhabitants (Badera and Kocoń, 2014).

At the same time, both localities share a strong history of resistance to mining. Participants of the climate camp identified local referenda as milestones in the struggle against coal, as their organisation and campaigning involved the public in the decision-making process about the prospects of coal mining. The size of the settlements, attendance, and questions posed in the referenda, differed in each case (see Table 1).

Local referenda held in Horní Jiřetín (CZ) and Babiak commune (PL) were both organised by municipal representatives and the results—in which more than 90% of attendees voted against the expansion of coal mining in both cases—became legally binding, as attendance reached the required quora. The referendum in Litvínov (CZ) was initiated by the local organisation *Kořeny* and did not meet the quorum necessary for a legally-binding result. Yet, 95% of attendees voted against coal mining expansion. Not only are the actual results of the referenda important, but the mobilisation of the public and collaboration with NGOs helped to establish personal connections between various actors. The importance of the referenda for overall decision-making processes has been rather limited, as the power of national authorities goes beyond the power of affected municipalities.

Table 1. Overview of local referenda against lignite mining in North Bohemian Coal Basin and Konin Basin.

Locality	Year	Attendance	Percentage voted against coal mining	Questions	Notes
Horní Jiřetín (CZ)	2005	75%	96%	Should the municipality use all legal measures to protect houses from demolition and inhabitants from eviction? Should municipal representatives enforce lignite mining limits as effective protective measures against state and regional institutions?	
Litvínov (CZ)	2006	38%	95%	Should the city of Litvínov actively use all means to prohibit lignite mining to come closer to the city, while breaching the lignite mining limits established by governmental resolution nr. 444/1991?	Did not meet the quorum
Babiak (PL)	2015	43%	90%	Are you for the construction of a lignite mine within the Babiak region based on the „Deby Szlacheckie" field?"	

Fig. 1. Front banner of the march leaving Klimakemp in 2018 reads: "Do not change the climate, change ourselves. End coal." Photograph courtesy of Petr Vrabec (2018).

Local resistance with tangible results, such as local referenda, creates a fertile ground in which new forms of resistance can then take root, especially when certain forms of injustice remain. Lignite combustion contributes enormously to carbon dioxide emissions and thus to climate change. Climate camps thus contest the causes of climate change in the very areas where fossil fuel extraction takes place and usually share several common objectives: 1) they are an open and horizontal, self-organised space; 2) they are a sustainable, prefigurative event; 3) they foster communication among people in the movement; 4) they provide a safe space near fossil infrastructure for organising direct action against it. Mass direct civil disobedience is one of the defining traits of the climate justice movement and provides a substantial point of differentiation between separate streams in the climate movement (Kenis and Mathijs, 2014). Direct action has moved from global climate summits to the ground-zero of fossil fuel extraction (de Moor, 2020). Unlike the goal of civil disobedient action during the summit—that is, to block and disrupt negotiations during formal meetings (Chatterton et al., 2013)—the goal of direct action during camps is to block the operation of the mining machines.

More than that, they provide a place to contest worldviews through the solidarity-based interactions on which they are literally built

(Kaufmann et al., 2019). Each climate camp is a site- and context-specific event, altering its name according to the local context. In the text, I refer to the Polish climate camp as *Obóz dla Klimatu*, and to the Czech one as *Klimakemp*.

Methodological Approach

This chapter is based on ethnography and participant observation. My participation in climate camps varied in terms of time and intensity, from being involved in the long-term preparation of the event, to participation in the event only. I understand my research approach as a 'militant' one (Russell, 2015; Urla and Helepololei, 2014), where I consider myself part of the political struggle, reflecting autoethnographic moments and co-producing knowledge together with research partners. Militant research might be criticised as failing on objectivity, due to ready bias on behalf of the preferences and political allegiances of the researcher. However, no research is produced in an apolitical vacuum. The pretense of objectivity in research might therefore reproduce and reinforce unjust status quos. Militant research may also fall short if it does not provide a salient critique of the movement it studies, even though it might create valuable knowledge for the movement (Halvorsen, 2015). Militancy could provide a novel point of view from within the movement, to help understand and position why anti-coal resistance matters for the wider climate justice movement. During this research I have conducted informal and semi-structured interviews, and I have also participated in the production of media outcomes created by the movement (Müller and Morton, 2018).

This chapter is a comparative inquiry as I focus on four events: climate camps organised in two consecutive years in two localities of lignite extraction. Gingrich and Fox (2002) emphasise the importance of an engaged approach in comparative anthropology that reflects power relations and personal trajectories. Comparative inquiry is based on engaged anthropological research in the vein of the classical environmental slogan "think global, act local" (Peacock, 2002). I also understand it as related to Giri's (2006) call for ontological epistemology, where multi-sited fieldwork is important for a comparative effort that acknowledges partial connections rather than amplifying wholesale systemic differences.

Environmental and climate justice scholarship is sometimes criticised for being too connected to the movement and particular case studies. For example, Jenkins (2018) sees climate justice as an all-encompassing and incoherent framework to be sufficiently translated to legislative procedures and mechanisms and thus not effectively delivering justice to affected social groups. In the following section, I focus on events where climate justice is formed as the political subject on the ground, where actions transgress resistance beyond geographic location and across scales. To conduct resistance against lignite mining includes political meaning and is part of "social processes, through which people make sense of their world" (Simmons and Smith, 2017). I follow the work of Carl Death (2010) on counter-conduct; that is, the moments of interaction with various layers of state power through which contentious subjectivity emerges. It is not only material and historical conditions that shape contentious interactions, and emotions also constitute the performativity of these actions (Houston and Pulido, 2002; Juris, 2008). In other words, acts of resistance matter for participants too. Contentious activities have a profound importance for a collective identity as it creates space for the cooperation of various social groups, where cooperation would have been hardly imaginable in earlier times (Polletta and Jasper, 2001).

Defying the Power around Lignite

Resistance to lignite mining was performed and articulated in different ways during protest events. Here I focus on the elements of the resistance that are disruptive to the power that enables and causes lignite mining to endure in an EE context. The existing power structures counter the resistance to lignite mining by deploying precautionary measures as part of a repressive apparatus against participants of climate camps. In turn, climate protests and activities attempt to overcome these silencing measures to disrupt the mining operation and the discourse that justifies it. Participants attempt to respond to the policies and legislation that influence energy metabolism.

Undoing the Extremist Label

Before the first climate camp in the Czech Republic in 2017, the Ministry of the Interior listed *Klimakemp* within their report in the section on 'Leftist Extremism'. As a response, the climate justice movement released an open letter addressed to the Prime Minister and Minister of the Interior, where not only the legitimacy of the listing but also the understanding of the term extremism was contested ("Otevřený dopis," 2017). In the open letter, the term 'extremism' was reinterpreted as a means of assuming a 'business-as-usual' approach to fossil fuel extraction. The letter also argued that when participants build and maintain a space for climate justice, they demonstrate that protection of the climate is not a crime, but part of the common, public interest. This counter-conduct to the imposed label of extremism could be seen as part of the creation of the subjectivity of the movement, as the action of the movement showed the fossil fuel industry as the real threat to society, rather than the peaceful protests (Kurik, 2016).

During the protest actions, police used the opportunity granted to them by precautionary measures against activists, to cause inconvenience and spread uncertainty. In 2019, police searched buses carrying protestors to the legally announced and permitted demonstration in Kleczew, checking the technical state of the buses, but also the personal information and identification of the passengers. The material aspects of precautionary and repressive measures also changed. Throughout the route of the demonstration at the Bílina mine in 2018, police accompanied protestors using various kinds of equipment, including horses, quads, a helicopter and other vehicles. Technological and power dominance was a clearly visible part of police conduct, such as when approximately half the protestors, who broke rank *en route* to the mine, were caught within a few metres. Despite their technological dominance, officers acted in a physically aggressive way, notably against a participant who suffered broken ribs, and to whom they did not provide immediate medical help.

Yet the 'precautionary measures' of the repressive apparatus also had protective consequences for the movement. The presence of the police hindered the threat of direct physical violence by opposition groups, which in the case of Poland were associated with the Facebook page *Odkrywkowo*. After the action, *Odkrywkowo* posted an acknowledgement of police work. On the one hand, they lamented the use of batons and tear gas against protestors, while, on the other, the post ended with the implication that without police presence, the protest could have ended up as a violent confrontation between miners and participants of the protest: "If there would be an open confrontation with miners—green extremists would collect financial contributions for other purposes..."—implying the need to cover medical expenses for those beaten.

The above-mentioned preventative and repressive measures show a part of what it takes to preserve the operations of lignite infrastructure. The costs of social resistance to mining operations can also be expressed financially, as for instance when the mining company *Severočeské doly* claimed an economic loss of more than 660,000 CZK (approximately €25,800) against several dozen activists. Economic loss calculations reflect how the mining company values mining operations, but also show the costs that direct action can cause. One of the central slogans of the movement— "We are the investment risk"—is financially expressed by the mining company and the claim could be perceived as an attempt to externalise the burden on the participants of the resistance. However, at the same time, the claim has become one of the mobilising arguments for the movement. The action of the movement has caused economic loss to what was perceived by protesters as illegitimate extractive activity, and thus sought to make less profitable. At the same time, the movement provided solidarity and support to prosecuted participants, which also served to attract new participants, who may have been scared of the costs and consequences of future actions.

Contesting Coal Politics: Protests as Reaction to Decision-Making

To question certain procedures publicly also highlights the disenfranchisement and lack of meaningful opportunities for participation in the extant procedures of ostensibly 'democratic' government. When, in 2015, the Czech government renegotiated a key resolution known as 'limits to lignite mining', the most extractive scenario incorporated the need to demolish the city of Horní Jiřetín (Lehotský and Černík, 2019). Nation-wide public mobilisation against this emerged around the slogan "*Limity jsme my*"—"we are the limits".

The slogan expressed a willingness to put protestors' own bodies in the place of insufficient governmental considerations. While governmental renegotiation of the position of 'limits' resulted in a victory for Horní Jiřetín, limits were still breached at a different open-cast mine, Bílina. Two years after the renegotiation, the slogan transferred to the movement that organised the first *Klimakemp* at the saved Horní Jiřetín. The camp questioned the broader consequences of lignite mining that were not appropriately taken into account during the renegotiation, particularly the mine's contributions to climate change.

The first *Obóz dla Klimatu* was organised in summer 2018, half a year before Poland hosted the UNFCCC's Conference of the Parties in Katowice. Polish dependency on coal was in the international spotlight and people resisting domestic lignite mining highlighted their cases amidst large international protests, notably the *March for Climate*. These resistances ruptured the image of Poland that its President Andrzej Duda had tried to claim (Berendt, 2018). The following summer, in 2019, the protest gained greater legitimacy as an EU infringement procedure was raised against Poland for continued mining operations at the Tomislawice mine. According to the European Commission, this mine was working without a valid hydrological permit ("Fundacja RT-ON—Komisja Europejska: odkrywka Tomisławice narusza prawo," n.d.). Protests during the second *Obóz dla Klimatu* attempted to block mining operations directly, when neither regional nor national authorities acted against the mine despite its infringement of legal requirements.

Building Counter-Hegemonic Imaginaries

In the previous section, we have seen how protests created ruptures in the power networks that sustain lignite extraction. However, the resistance also creates ground for new demands. Protests connect various actors and open up possibilities for their future cooperation and for the building of alternatives. In this section, I will provide examples of the construction of counter-hegemonic alternatives during protest. First, this relates to internationalisation, whereby symbolic means and tangible acts connect local struggles against lignite mining with the global climate justice movement. Second, I will focus on the acts connecting previous struggles at the sites with the emerging climate justice movement. Finally, I elaborate on moments in which various contemporary social and emancipatory struggles have converged with the ecological.

Internationalisation

Climate camps appropriate internationally recognised symbolic and visual instruments from the global climate justice movement and translate them into a local context. During the *Klimakemp* in 2017, participants formed a red line, joining hands in a human chain at the edge of the Czechoslovak Army Mine with a red line sprayed along the sleeves of their white overalls. Chanting "*Limity jsme my*", the human red line embodied a governmental resolution. A red line was also formed during "picketing"—showing statements on banners and pickets—on the first *Obóz dla Klimatu,* serving as an object for cameras and media images, with the Patnow powerplant in the background. The year after, another red line served as a meeting point for all previously separated action groups as participants met again in the park in Krzyszkowice, where they started to spread alongside the road on the edge of the Tomislawice mine, going through demolished parts of the village. The large red ribbon stretched several hundred metres, connecting groups of participants previously separated by different action levels (those engaged in civil disobedient action and those who were not).

Similar to the red line, white overalls became a symbol of the climate justice movement in Europe, notably in Ende Gelände actions in Germany (Sander, 2017). They contrast with black bloc tactics, bearing a different meaning and historical legacy of non-violence (Juris, 2008). In these civil disobedient actions, white overalls created a collective body, as individuals were hardly distinguishable from each other in the crowd. Along with respiratory masks, white overalls provide protection from dusty environments, but also make it harder for police to act against the uniformed crowd and to identify each participant.

International participants are directly involved in activities at all climate camps. At *Obóz dla Klimatu* in 2018, anti-coal activists from Colombia shared their very different experiences of struggles: first in a workshop, and then, during the protest, one of the activists became a spokesperson to the media. When replying to a TV reporter's question about why she does not protest at home, she articulated the protest as part of the struggle for Mother Earth ("*Protest ekologów przed konińską elektrownią,*" n.d.). The slogan, which translates as "Enough compromises in the protection of Mother Earth", provides a counterpoint to the nationalist discourse on energy security based on domestic lignite production and expands the notion of citizenship—not only citizens or locals become seen as having the right to intervene in the domestic regime of extraction.

Climate camps also provide opportunity and inspiration to develop the struggle. The idea of organising a climate camp in Poland was raised during the first climate camp in the Czech Republic. As one participant describes it: "I was in *Limity* in the Czech Republic last year, with two other Polish people who are also in the organisation group. We were sitting, drinking beers and thinking: Hey, let's do something like this in Poland next year," (personal interview, 2018). This is what Tormos-Aponte and García-López (2018) describe as mutual learning in the movement, bringing recognition of the struggle to other localities.

Building on Previous Struggles

If we understand the building of international connections as horizontal, with relations to the history of previous and future resistance forms as interconnected, then there may also exist vertical connections located in particular places. In both regions, across Poland and the Czech Republic, a history of struggle against lignite mining exists, notably conducted by local inhabitants against its direct effects. Climate camps enable people who are also affected by other aspects of lignite mining to find common ground within this heritage.

By 2015, there were various forms of resistance ranging from referenda to a tractor blockade against the plans of the mining company, ZE-PAK, to expand open cast mining in the region. Protests highlighted the exploitative nature of the enterprise, often personified by the owner of the company, Zygmunt Solorz-Zak. Pickets, banners and slogans addressed to the owner, such as "Mr. Zygmunt, not on our land" also appeared at the climate camp. Although not all the participants and organisers of the previous protests attended the climate camp in person, some of the representatives of the initiatives joined climate camps and held speeches during protests at *Obóz dla Klimatu*.

During the 2019 *Klimakemp*, connections to previous struggles were made directly during the protests. On the march, for instance, protestors walked through Libkovice, the last village demolished because of lignite mining in 1993. At that time, civil disobedience blockades attempted to stop the demolition and the actions were some of the first by the emerging environmental movement in the Czech Republic. Participants of the original blockade also joined the march in 2019 and spoke about their experiences at the spot where the first round of speeches had taken place. Through such moments, the resistance of the past was felt in the present, commemorating the history of the environmental movement and resistance against coal mining in the country.

Convergence Nodes of Emancipatory Struggles

Apart from historical and regional struggles, current social resistance also creates conditions in which climate justice movements emerge. Connections between various emancipatory struggles form an imaginary of viable alternatives to the injustice connected with lignite mining. The emerging climate justice movement in Poland was founded from the very beginning on delicately balanced cooperation between various groups, including local farmers, anarchist groups and NGOs. *Obóz dla Klimatu* in 2018 set a goal to create a safe space, a laboratory where different groups of people can meet in a respectful atmosphere to create an easily accessible entrance point into the movement for new participants and, to that end, refrained from civil disobedient action.

During the protests at *Klimakemp* and *Obóz dla Klimatu* significant solidarity was expressed with workers of the lignite mining companies. Protesters brought their translation of the just transition, which was originally a unionist concept (Stevis and Felli, 2015). The expression of solidarity with workers was evident through slogans in the protests, such as "*in solidarity with workers, but never with open-pit mines*", or "*four-day working week*" as a rather desirable, one-sided plea. Employees or union members of the mining companies did not participate in the debates organised during the camp's programme, despite their invitation. After the first *Obóz dla Klimatu*, the concept of just transition was contested by *Inicjatyva Pracownicza*, an independent unionist magazine in Poland (Urbanski, 2018). Both *Obóz dla Klimatu* and *Klimakemp* attempted to discuss democratisation of the energy sector. Common ground for an imaginary beyond coal mining, that included both workers of the mining industry and the climate justice movement, remained elusive both inside and outside of climate camps. Currently both regions are involved in the EU Platform for Coal Regions in Transition and it is questionable whether this platform will provide an opportunity to move beyond an approach to just transition based on a 'green economy'.

Critique of green growth or market-based solutions is contentious in the environmental movements of post-communist countries, where liberal environmentalism has been prevalent (Jehlička et al., 2005; Novák, 2015). However, groups identifying themselves as anti-capitalist also participate at climate camps. The Polish slogan "People first, profits later", or the front banner at the *Klimakemp* 2018 saying "Don't change the climate, change ourselves. End coal", emphasise the importance of systemic critiques articulated during protests. Rather than silencing anti-capitalist voices, careful translations of the overarching central slogan of the climate justice movement—"System Change, Not Climate Change"—situate the systemic critique and vocabulary of these movements in Poland and the Czech Republic.

Various voices are present in climate justice resistance in the regions of lignite extraction, some of which might not be perceived as directly connected to the issue of lignite mining and climate change. After their return from the protest to the 2019 camp at *Obóz dla Klimatu*, participants learned about violent attacks on the Bialystok LGBT Pride March. Shocked and shaken, participants immediately organised a solidarity group picture, taking a clear stand for the value of diversity—a controversial topic in contemporary Poland. Such tangible acts of inter-movement spill-over serve to broaden the realm of justice (Hadden, 2014).

Conclusion

Climate camps in the Czech Republic and Poland have politicised the dependency of both states on coal and the extraction of lignite. Through resistance, the area of extraction becomes a place of contestation, but also becomes connected and intertwined with the global scale of the

Fig. 2. Group picture at Obóz dla Klimatu in solidarity with Pride in Bialystok. Earlier that day, on 20 July 2019, the Pride march was violently attacked. Photograph courtesy of Ewa Bielańczyk-Obst (2019).

climate crisis. Resistance is not a monolithic activity—each event is situated in specific conditions that make it significant. In this chapter, I have identified two aspects of the resistance: disruptive and constructive. I have looked at certain moments that made resistance significant for the participants, but also for the opponents of the climate justice movement.

The disruptive force of resistance is evident in the precautionary and repressive measures deployed against the protestors before, during, and after climate camps. From discursive efforts, such as framing the resistance as extremist, to the technological dominance of police during the protests, or the calculation of the economic loss caused by the protests—all of these tactics demonstrate a concerted effort to contain the disruption caused by the resistance. Such resistance is also situated within the political and legislative context that influences lignite mining. Protestors politicised specific government resolutions in the case of lignite mining limits in the North Bohemian Coal Basin and the infringement of water permits in the Tomislawice mine, in Poland, or disrupted the national discourse of energy security based on domestic coal production.

Climate camps in Poland and the Czech Republic are also examples of constructive resistance. In these camps local cases are symbolically bound to the global and international context, with visual means of sharing amongst members of the global climate justice movement being deployed. Climate camps are open to international participation and thus expand energy citizenship beyond the nation-state. The camps also acknowledge and build on previous struggles, introducing knowledge about the previous protests to new participants and expanding their awareness of contemporary issues of climate justice. The resistance at climate camps also broadens the realm of justice, which might not be perceived as directly related to coal mining or climate change. It opens up debates about the democratic imaginaries of nature-society relationships within the movement, for example by contesting the meaning of any singular 'just transition'.

Although we can understand climate camps in both countries as events of the same global climate justice movement, in this chapter I have also interrogated important differences in the translation of tactics and organisation of camps to their respective contexts. The movement reacts to broader social problems to challenge hegemonic power. Probably the most notable difference between the Czech and Polish cases emerged in Poland, where climate resistance was extended to encompass support for the LGBT Pride march in Bialystok. This was in response to violent threats from those opposing the messages of the climate camp. Such broad social- and identity-entanglements have not as yet been a characteristic of the Czech environmental resistance context.

Climate camps in both Konin and the North Bohemian Coal Basin create a leverage point that pushes for an earlier coal phase-out in their respective states, both of which are considered part of the coal heartland of the EU. These events matter to participants as they empower them to participate meaningfully in shaping society's energy metabolism by intervening in the dynamics of lignite mining. This resistance to lignite mining at climate camps seeks to rupture a socio-technological imaginary based on coal dependency. It also sparks a debate about broader socio-ecological transformations, politicising the issue of justice, and opening up new possibilities of a post-carbon future. More than a decade after the first climate camp in Britain, *Klimakemp* and *Obóz dla Klimatu* show how this way of organising has travelled and expanded to new regions and has been adjusted to EE contexts.

References

Andersson, M., 1999. Change and Continuity in Poland's Environmental Policy, Environment & Policy. Springer Netherlands, Dordrecht. https://doi.org/10.1007/978-94-011-4517-6.

Badera, J., Kocoń, P., 2014. Local community opinions regarding the socio-environmental aspects of lignite surface mining: Experiences from central Poland. *Energy Policy*, 66, 507–16. https://doi.org/10.1016/j.enpol.2013.11.048.

Bergman, N., 2014. Climate Camp and public discourse of climate change in the UK. Carbon Management, 5, 339–48. https://doi.org/10.1080/17583004.2014.995407.

Brown, B., Spiegel, S. J., 2019. Coal, Climate Justice, and the Cultural Politics of Energy Transition. *Global Environmental Politics*, 19, 149–68. https://doi.org/10.1162/glep_a_00501.

Chatterton, P., Featherstone, D., Routledge, P., 2013. Articulating Climate Justice in Copenhagen: Antagonism, the Commons, and Solidarity. *Antipode*, 45, 602–20. https://doi.org/10.1111/j.1467-8330.2012.01025.x.

Císař, O., 2010. Externally sponsored contention: the channelling of environmental movement organisations in the Czech Republic after the fall of Communism. *Environmental Politics*, 19, 736–55. https://doi.org/10.1080/09644016.2010.508305.

D'Alisa, G., Demaria, F., Cattaneo, C., 2013. Civil and uncivil actors for a degrowth society. *Journal of Civil Society*, 9, 212–24. https://doi.org/10.1080/17448689.2013.788935.

de Moor, J., 2020. Alternative globalities? Climatization processes and the climate movement beyond COPs. *International Politics* (2020). https://doi.org/10.1057/s41311-020-00222-y.

Death, C., 2010. Counter-conducts: A Foucauldian Analytics of Protest. *Social Movement Studies*, 9, 235–51. https://doi.org/10.1080/14742837.2010.493655.

Energy Transition in Poland — Forum Energii [WWW Document], n.d. URL http://forum-energii.eu/en/polska-transformacja-energetyczna.

ERÚ, 2018. *Roční zpráva o provozu ES ČR 2017*. Energetický regulační úřad, Praha.

Flisowska, J., Moore, C., 2019. *Just Transition or Just Talk?* Climate Action Network Europe. https://caneurope.org/content/uploads/2019/05/Just-Transition-or-Just-Talk-Report.pdf.

Frantál, B., Nováková, E., 2014. A curse of coal? Exploring unintended regional consequences of coal energy in the Czech Republic. *Moravian Geographical Reports*, 22, 55–65. https://doi.org/10.2478/mgr-2014-0012.

Frenzel, F., Feigenbaum, A., McCurdy, P., 2014. Protest Camps: An Emerging Field of Social Movement Research. *Socioligcal Review*, 62, 457–74. https://doi.org/10.1111/1467-954X.12111.

Fundacja RT-ON — Komisja Europejska: odkrywka Tomisławice narusza prawo [WWW Document], n.d. URL https://rozwojtak-odkrywkinie.pl/component/k2/item/363-komisja-europejska-tomislawice-naruszaja-prawo.

Gingrich, A., Fox, R. G., 2002. Introduction, in: Gingrich, A., Fox, R. G. (Eds.), *Anthropology, by Comparison*. Routledge, London and New York, pp. 1–22. https://doi.org/10.4324/9780203463901.

Giri, A. K., 2006. Creative Social Research: Rethinking Theories and Methods and the Calling of an Ontological Epistemology of Participation. *Dialectical Anthropology*, 30, 227–71. https://doi.org/10.1007/s10624-007-9007-8.

Hadden, J., 2014. Explaining Variation in Transnational Climate Change Activism: The Role of Inter-Movement Spillover. *Global Environmental Politics*, 14, 7–25. https://doi.org/10.1162/GLEP_a_00225.

Halvorsen, S., 2015. Militant research against-and-beyond itself: critical perspectives from the university and Occupy London. *Area*, 47, 466–72. https://doi.org/10.1111/area.12221.

Hollander, J. A., Einwohner, R. L., 2004. Conceptualizing Resistance. *Sociological Forum*, 19, 533–54. https://doi.org/10.1007/s11206-004-0694-5.

Houston, D., Pulido, L., 2002. The Work of Performativity: Staging Social Justice at the University of Southern California. *Environment and Planning: Society and Space*, 20, 401–24. https://doi.org/10.1068/d344.

Jehlička, P., Sarre, P., Podoba, J., 2005. The Czech Environmental Movement's Knowledge Interests in the 1990s: Compatibility of Western Influences with pre-1989 Perspectives. *Environmental Politics*, 14, 64–82. https://doi.org/10.1080/0964401042000310187.

Jenkins, K., 2018. Setting energy justice apart from the crowd: Lessons from environmental and climate justice. *Energy Research & Social Science*, 39, 117–21. https://doi.org/10.1016/j.erss.2017.11.015.

Juris, J. S., 2008. Performing politics: Image, embodiment, and affective solidarity during anti-corporate globalization protests. *Ethnography*, 9, 61–97. https://doi.org/10.1177/1466138108088949.

Kaufmann, N., Sanders, C., Wortmann, J., 2019. Building new foundations: the future of education from a degrowth perspective. *Sustainability Science*, 14, 931–41. https://doi.org/10.1007/s11625-019-00699-4.

Kenis, A., Mathijs, E., 2014. Climate change and post-politics: Repoliticizing the present by imagining the future? *Geoforum*, 52, 148–56. https://doi.org/10.1016/j.geoforum.2014.01.009.

Klein, N., 2015. *This Changes Everything: Capitalism vs. The Climate*, 2nd edition. Simon & Schuster, New York.

Kuchler, M., Bridge, G., 2018. Down the black hole: Sustaining national socio-technical imaginaries of coal in Poland. *Energy Research & Social Science*, 41, 136–47. https://doi.org/10.1016/j.erss.2018.04.014.

Kurik, B., 2016. Emerging Subjectivity in Protest, in: *The SAGE Handbook of Resistance*. SAGE, London. https://doi.org/10.4135/9781473957947.n3.

Kuskova, P., Gingrich, S., Krausmann, F., 2008. Long term changes in social metabolism and land use in Czechoslovakia, 1830–2000: an energy transition under changing political regimes. *Ecological Economics*, 394–407. http://dx.doi.org/10.1016/j.ecolecon.2008.04.006.

Lehotský, L., Černík, M., 2019. Brown coal mining in the Czech Republic — lessons on the coal phase-out. *International Issues: Slovak Foreign Policy Affairs*, XXVIII, 45–63.

Lloyd, M., 2009. Performing radical democracy, in: Little, A. and Lloyd, M. (Eds.), *The Politics of Radical Democracy*. Edinburgh University Press, Edinburgh. https://doi.org/10.3366/edinburgh/9780748633999.003.0003.

Macuk, R., Maćkowiak-Pandera, J., Gawlikowska-Fyk, A., Rubczyński, A., 2019. *Energy Transition in Poland*. Forum Energii, Warszawa.

Mazurek, S., Tymiński, M., 2019. Brown Coal 2019 [WWW Document]. Polish Geological Institute, http://geoportal.pgi.gov.pl/surowce/energetyczne/wegiel_brunatny (n.d).

Ministerstwo Energii, 2018. *Program dla sektora górnictwa węgla brunatnego w Polsce*.

Ministry of Energy, 2018. *Energy Policy of Poland until 2040 (EPP2040)* (Extract from draft). Ministry of Energy, Warsaw.

MPO, 2014. *State Energy Policy of the Czech Republic*. Ministry of Industry and Trade, Prague.

Müller, K., Morton, T., 2018. At the German coalface: Interdisciplinary collaboration between anthropology and journalism. *Energy Research & Social Science*, Special Issue on the Problems of Methods in Climate and Energy Research, 45, 134–43. https://doi.org/10.1016/j.erss.2018.06.016.

Novák, A., 2015. *Tmavozelený svět. Radikálně ekologické aktivity v České republice po roce 1989*. SLON, Praha.

Osička, J., Kemmerzell, J., Zoll, M., Lehotský, L., Černoch, F., Knodt, M., 2020. What's next for the European coal heartland? Exploring the future of coal as presented in German, Polish and Czech press. *Energy Research & Social Science*, 61, 101316. https://doi.org/10.1016/j.erss.2019.101316.

Otevřený dopis: ochrana klimatu není extremismus, 2017. *Limity Jsme My*. https://limityjsmemy.cz/2017/06/otevreny-dopis-ochrana-klimatu-neni-extremismus/.

Polletta, F., Jasper, J. M., 2001. Collective Identity and Social Movements. *Annual Review of Sociology*, 27, 283–305. https://doi.org/10.1146/annurev.soc.27.1.283.

Protest ekologów przed konińską elektrownią, n.d. https://poznan.tvp.pl/38162238/protest-ekologow-przed-koninska-elektrownia.

Rancière, J., 1999. *Disagreement: Politics and Philosophy*. University of Minnesota Press, Minneapolis, MN.

Rasch, E. D., Köhne, M., 2016. Hydraulic fracturing, energy transition and political engagement in the Netherlands: The energetics of citizenship. *Energy Research & Social Science*, Energy Transitions in Europe: Emerging Challenges, Innovative Approaches, and Possible Solutions, 13, 106–15. https://doi.org/10.1016/j.erss.2015.12.014.

Reitan, R., Gibson, S., 2012. Climate Change or Social Change? Environmental and Leftist Praxis and Participatory Action Research. *Globalizations*, 9, 395–410. https://doi.org/10.1080/14747731.2012.680735.

Russell, B., 2015. Beyond activism/academia: militant research and the radical climate and climate justice movement(s). *Area*, 47, 222–29. https://doi.org/10.1111/area.12086.

Russell, B., Schlembach, R., Lear, B., 2017. Carry on camping? The British Camp for Climate Action as a political refrain, in: *Protest Camps in International Context: Spaces, Infrastructures and Media of Resistance*. Policy Press, Bristol, pp. 147–62. https://doi.org/10.1332/policypress/9781447329411.003.0009.

Sander, H., 2017. Ende Gelände: Anti-Kohle-Proteste in Deutschland. *Forschungsjournal Soziale Bewegungen*, 30, 26–36. https://doi.org/10.1515/fjsb-2017-0004.

Saunders, C., Price, S., 2009. One person's eu-topia, another's hell: Climate Camp as a heterotopia. *Environmental Politics*, 18, 117–22. https://doi.org/10.1080/09644010802624850.

Severočeské doly a.s. [WWW Document], n.d. https://www.sdas.cz/aktivity/hornicka-cinnost/doly-bilina.aspx.

Simmons, E. S., Smith, N. R., 2017. Comparison with an Ethnographic Sensibility. *PS Political Science & Politics*, 50, 126–30. https://doi.org/10.1017/S1049096516002286.

Stevis, D., Felli, R., 2015. Global labour unions and just transition to a green economy. *International Environmental Agreements: Politics, Law, and Economics*, 15, 29–43. https://doi.org/10.1007/s10784-014-9266-1.

Temper, L., Demaria, F., Scheidel, A., Bene, D. D., Martinez-Alier, J., 2018. The Global Environmental Justice Atlas (EJAtlas): ecological distribution conflicts as forces for sustainability. *Sustainability Science*, 13, 573–84. https://doi.org/10.1007/s11625-018-0563-4.

Tormos-Aponte, F., García-López, G. A., 2018. Polycentric struggles: The experience of the global climate justice movement. *Environmental Policy and Governance*, 28, 284–94. https://doi.org/10.1002/eet.1815.

Urbanski, J., 2018. Niesprawiedliwa transformacja. Ekologia kontra gornictwo. *Biuletyn Zwiazkowy Inicjatywa Pracownicza*, 8–9.

Urla, J., Helepololei, J., 2014. The Ethnography of Resistance Then and Now: On Thickness and Activist Engagement in the Twenty-First Century. *History and Anthropology*, 25, 431–51. https://doi.org/10.1080/02757206.2014.930456.

4. Gaps of Warsaw
Urban Environmentalism through Green Interstices

Jana Hrckova

When Warsovians are asked to reflect on their city, they eventually steer the conversation towards the phenomenon that I call urban gaps. As Joanna Kusiak contends within the opening pages of *Chasing Warsaw* (Kusiak, 2005), the city is marked by many small and large open spaces with an interstitial character that are seemingly 'left-over' and derelict. This is a result of a combination of the planned and the unplanned, as Warsaw has a scattered structure of buildings stemming from modernist planning and the after-effects of the destruction wrought by WWII, as well as more contemporary development. As an illustration, the reader can imagine terrains that span from halted construction lots and squares turned into improvised parking lots, through to meadows and poorly kept lawns by the roadside, to dilapidated socialist sport complexes. As many of these hard-to-define areas bustle with vagabond greenery and can still be found in the very centre of the city, both the municipal government and developers often have their eyes on them as sites of lucrative future urban development.

Looking at these green spaces through an ethnographic lens, this chapter analyses shifts in urban policies that have led from formal stakeholders trying to undermine the social life in such areas, towards considering the ways in which these areas are today providing openings for a re-evaluation of the city and their role within it through environmental considerations. The chapter begins with an exploration of city hall policies as well as the prevalent imaginaries of neoliberal

 https://doi.org/10.11647/OBP.0244.04

Fig. 1. Empty area in Mokotow district in the vicinity of a housing estate about to be developed. Photograph by the author (2017).

spatial order and urban life, which have paradoxically often led to both the production and disappearance of the gaps, as well as to their dormant character. The analysis links such processes with norms associated with the fetishisation of private property in post-socialist countries (Hann, 2005). Sonia Hirt (2012) once called this phenomenon a condition of privatism, signalled by a disintegration of public spaces and a belief that the private realm ensures a thriving society. Besides the centrality of private property, the chapter focuses on both physical and metaphorical 'clean ups' within urban space that have been facilitated by the city hall, which have made the existence of these 'unruly' green gaps in the city undesirable.

On the other hand, zooming in on a contrarian approach to these gaps, the text also highlights the often unforeseen social value held by the urban interstices in Warsaw and their unique liminal condition in urban space (Berger, 2006: 31). Using an example of the Jazdow initiative as one such rebellious 'gap' in the city's very centre, the chapter examines the dynamic changes of the discourses of the municipal government in post-Global Financial Crisis (GFC) Warsaw and explores how the shifting official imaginaries as they relate to environmentalism and value have been used to 'save' some of these spaces. Seizing the re-opened possibilities of city hall policy through the creative city paradigm, the activists at Jazdow subversively employed the official language of participation and public space 'activation' to keep the area open for alternative social experimentation.

Here, I use the term 'activation' as an emic term that urban activists in Poland have used, denoting an opening up of spaces for public use so that they can be filled with all sorts of social life. Urban activation thus perfectly fits with newly embraced 'creative city' approaches, influenced by the now infamous works of urban consultants Richard Florida (2005) and Charles Landry (2000). Creative city development preaches that cultural resources and the symbolic content of urban economies ought to be harnessed, and "local peculiarities" for economic development appropriated (Novy and Colomb, 2013: 1821). The recent emergence of environmental concerns has also been used as a way to articulate urban struggles, to link up localised issues with national and global ones. While urban agendas are often perceived as confined to a specific place, embedding the conflict within transnational environmental activism has provided a boost in legitimacy as well as equipping activists with new tools in their work.[1] For example, the air pollution crisis that has resonated amongst the local population in Warsaw has allowed for a re-evaluation of what urban 'gaps' represent and could be as an inherent part of the urban realm, possibly putting an end to authorities and developers seeing them as black holes in the city.

A leftist, Warsaw-based publisher that I talked to about what I perceived as a lack of social life in the urban interstices of Warsaw exclaimed that "Warsaw is no Barcelona!" Visibly surprised by my problematising and questioning the processes that might be behind a 'no Barcelona' vibe, she continued: "People in Warsaw always make their own rules. When the need arises, they will figure something out." I therefore hope that the chapter will reveal that 'figuring something out'

1 See for example Timothy Choy's (2011) discussion on the centrality of striking the right balance between local and international expertise in environmental struggles in Hong Kong.

in relation to the green interstices and formulating possible alternatives for their future has already begun.

This chapter draws on ethnographic fieldwork that took place between 2015 and 2018. I conducted a series of semi-structured interviews with stakeholders ranging from activists at Jazdow to members of other urban movements, officials and academics. Further data was acquired from participant observation at Jazdow and in various other gaps as well as a qualitative analysis of policy and media outputs. During my fieldwork, I was an active member of a community gardening group at Jazdow.

Fig. 2. Jazdow community garden. Photograph by the author (2018).

The Social Life of 'Gaps'

I have derived the 'gaps' concept from the emic way in which Warsaw urban interstices are discussed amongst people: these places are often referred to as empty or desolate. In this chapter, such gaps will be used as an umbrella term for the informal, interstitial spaces that often remain from wartime destruction or poor planning, spaces that either lie derelict or are not used for the purposes for which they were designed.

In eastern Europe (EE), and in Warsaw in particular, the urban interstices have a special significance due to particular histories of destruction and consequent waves of (sometimes unfinished) re-builds and constructions that have contributed towards the spatial organisation of the city. If these gaps appear at all in the literature, they are identified as being the result of post-industrial economies giving rise to abandoned factories in the inner cities (Edensor, 2005) or of urban sprawl (Sievert's *Zwischenstadt* concept, 2001). In Warsaw such spaces have been present since the destruction of WWII (when approximately 84% of the city was destroyed) and perpetuated by the numerous, often unfinished construction projects that took place on the outskirts. Zydek (2014: 76) points out that "Warsaw's emptiness has often had a planned character," as the city outline included monumental open spaces that have been hard to maintain, and often originally served predominantly ideological goals, such as the huge Defilad Square in the city centre that hosted regular parades. The urban development of the city was fully controlled by planners, as in 1945, 93% of land in the capital was nationalised by the so-called Beirut Decree (Kusiak, 2012: 303). The move provided the communist party with a free hand in deciding spatial development, stripping the land of its commercial value and opening possibilities for an airy, modernist development that extended beyond the city centre.

The disintegration of the socialist regimes ushered in further 'gaps' that appeared during the 1990s together with the destruction of public facilities and infrastructure, and sprawling suburbanisation. Urban interstices must be placed in the context of the post-socialist reality and "actually existing neoliberalism" (Brenner and Theodore, 2002) that swept through the whole EE region from the 1990s. The processes that neoliberal governance unleashed led to an interesting paradox. While the urban interstices started disappearing quickly during the late 2000s, as they filled up with new constructions, others paradoxically sprang up in spaces where busy urban life used to take place. These new spaces arose due to failing public infrastructure and ownership irregularities. The fuzziness of actual ownership and the messy land restitution process contributed to the production of these new gaps. As Murawski (2018: 28) shows, fuzzy property issues in

Warsaw have been severely detrimental for ordinary inhabitants,[2] as they have led to the further exploitation of the already dispossessed while the wealthy and corrupt have benefited from this fuzziness by being able to manipulate ownership restitution titles and processes in their favour.

Scholars of political ecology have demonstrated the stark economic and developmental unevenness that can be found in urban landscapes (Heynen, 2014); in Warsaw such unevenness can be traced in the natural and social lives of gaps as well as in the gaps that no longer exist. Interstitial spaces are thus today infused with meaning stemming from such histories, serving as some of the most important sources of contemporary traces of the pre-war, socialist, but also recent capitalist past, through which unique ecologies are produced. Despite often being framed in moral dimensions of waste and decay, remaining urban gaps have been perceived to stand out by authorities and developers from value-producing structures of the city (De-Sola Morales, 1995: 120). They can be researched as a mitigating border zone, presenting limits to the omnipresent commodification of space, as well as acting as a last refuge and potential for alternative uses. They can be seen as places that are not "stage-managed for tourism or consumption" (Gandy, 2009: 152), or as places with ruderal ecologies where unexpected human and non-human neighbours can come together.

Upon entering my field, and influenced by the existing literature on gaps, vague and left-over spaces, I assumed that I was to find such spaces filled with all sorts of social life that had been squeezed out of the mainstream and increasingly uneven Warsaw (Berger, 2006: 31; Saksouk-Sasso, 2013; Anderson, 2015; Tonnelat, 2008). I expected to find tiny markets there; I expected to find people, who used these spaces as restrooms, meeting places, spots to take their dog out or sell crack, as sleepover areas, sites for game scenarios for kids or hipster hangouts, the world of the subaltern and the world that exists beyond the rules of the profit-oriented capitalist city. To my great surprise, however, I found that not much was happening in these so-called 'gaps', in a social sense. In general, informants agreed with this observation, but the range of

2 As opposed to Verdery's description of fuzzy property in Romania (1998: 179), which sometimes allowed people to manipulate and navigate new sets of rules and relations to their favour.

their interpretations has been a challenge to process. Some attributed the lack of social action to cold weather during the first months of my research in late autumn 2016. One man I talked to even suggested that the oft-unemployed working class whom he met while hanging out in such spaces simply "moved to England" after accession to the EU.

I believe that underlying causes of such dormancy could be clearly identified in the largely unchallenged mode of neoliberal governance that has functioned in the city for almost two decades. After the shock therapy reforms in the early 1990s and a subsequent economic slump that aided the creation of many of the gaps in the city, it was the 2000s that brought in a new wave of financialised corporate capital. Poblocki suggests that this was the time when earlier Polish "car boot sale" capitalism swiftly changed, accompanied by a construction boom (Poblocki, 2012: 272). The scale of development led to more stringent execution of property rights, making it more difficult to appropriate spaces with known owners. It is also vital to understand the inherent desire of the municipality and some of the inhabitants at this time to show that Poland was a 'normal'[3] country, and to mimic the perceived ways of being Western and modern.

'Hoping for normal' in this sense can be linked to the ideology of post-socialist transition that understands the changes in the 1990s as a trajectory of catching up, moving back (or forward) towards democracy, capitalism and civilisation (Stenning and Hörschelmann, 2008: 321). Unsurprisingly, the prevalent imaginary of the desired spirit of Warsaw[4] mirrored the capitalist ideals of the late 2000s and 2010s, attempting to model the city as a sterile safe haven for capital as well as for the upper classes. In the case of the gaps, the 'normal' stands for a well-landscaped and maintained city, where the shabbiness of both the urban space and its inhabitants would be relegated to the socialist past, to showcase the 'winners' in the new economy.

3 The term often used among my informants when describing their desires and ideas about the future or notions of a 'good life'. Similar use of the term 'normal' in post-socialist urban space has been recorded by Fehervary (2002), who wrote about how conspicuous consumption in the households in Dunaujvaros would be interpreted as returning to the wished-for normal; or Jansen (2015), who described the yearning of the Sarajevans for a working and predictable state.

4 A nod to Bell and Shalit's *Spirit of Cities* (2014).

In the late 2000s, the city hall unleashed a process of what could be called a 'beautification' of Warsaw. This meant not only making the city cleaner of trash and 'unkept' greenery, but also cleaning its public spaces of unwanted individuals. The ban on public drinking that has been in place since the 1980s curfew became more strictly policed. Similarly, small illegal market stalls were targeted and many legal ones had their permits removed. A friend made a comment that Warsaw from this time had the 'clean' aura that cities have under authoritarian rule—a cleanliness that is heavily enforced. The inspiration for such measures could be traced to Rudi Giuliani's infamous clean-up of the streets of New York and his introduction of 'broken window theory', which eventually became a playbook for mayors around the world wishing to market their "world class cities" (Schindler, 2014). An urban renaissance desired by some resulted in the displacement and harassment of others, and Warsaw was no different.

The often-dormant character of the gaps can also be attributed to a widespread post-socialist emphasis on the centrality of private property in everyday life and as an integral part of society (Hann, 2005: 550). Linking to the aforementioned fuzzy property issues in Warsaw, it is important to understand the double-edged consequences of the fetishisation of private property—on the one hand, it led to the emergence of gaps, as the rights and titles to lots in the city were often unclear, such that formal development was kept at bay, while at the same time, such uncertainty prevented social life within them, as such areas became fervently policed and thus untouchable by anyone save the (often unknown) owners.

The resulting relative abandonment of the gaps gave rise to a very convenient situation for real estate developers, aided by the city hall, who depicted many of the gaps as useless, dangerous and ugly, as areas with shrubbery standing in the way of the much desired 'densification' and a more compact city.[5] Warsaw has an unusually diffuse built structure, causing problems with the provision of public facilities and public transport and effectively enhancing car usage amongst some inhabitants. However, the quest for urban density has been used to justify

5 There have been numerous events (e.g. the Dense City conference held in January 2017 and the Dense City Workshop in February and March 2017) hosted by the city hall to discuss the options for further construction on unused urban terrain.

all kinds of projects, often destroying green spaces without responding to the above-mentioned challenges.[6] The approach has led to decades of "wild development," as inhabitants derogatively describe it, enabled by a virtual lack of binding planning processes in Warsaw, which has turned the city into a developers' playground. As an illustration, the number of new developments in 2008 was equal to all new constructions in London in the same year, a city four times the size of Warsaw (Kusiak, 2012: 302). While such a scenario was to be expected on privately-owned lots, the construction and privatisation fever has also frequently affected publicly-owned land. In the following sections, I will explore a strategy of resistance employed by the urban activists devoted to preserving one such publicly-owned space.

Gap Struggles

Here, I zoom in on Jazdow as one of the few cases where urban gaps have been preserved from real-estate development (at least for the moment) through the concerted action of local activists and inhabitants. I use this case to explore how urban imaginaries of city hall authorities were 'opened up' and redefined, where desired forms of urbanity could become a manoeuvring space for the advocates of the preservation of Jazdow.

The Jazdow settlement has an intriguing history: it arose in the very centre of the city as almost one hundred small wooden houses that were sent to Poland by the Finnish government as part of WWII reparations. The area is today approximately 5 hectares and full of lush greenery. The houses are municipally owned and were conceived as a temporary solution amidst the post-war housing shortage. Out of the original ninety houses, about twenty-six still stood during my research in 2017. Some of them were empty, some hosted the remaining twenty-five permanent inhabitants (as the city stopped allotting the houses and the former inhabitants left or died) and some were occupied by various NGOs and

6 There are numerous examples of questionable development: for example an (ongoing) attempt to start construction in the middle of the Bemowo Forest at a lot between two conservation areas. Similarly, an environmentally diverse area popular among Warsovians around Czerniakowski Lake has been picked for future development.

activists. The settlement has undergone a slow dismantling ever since the 1970s, however it took until around 2010 for the district authorities to proceed towards a complete clearing out of the area. The district mayor (representing the centre-liberal party that, in 2020, still governs city hall) argued that having a 'village' in the city centre is a shame, as the area was way too valuable to host overgrown shrubbery and a few shabby houses. The area truly lies within the most exclusive zones in the city, both in terms of prestige and land prices. The Polish Parliament is on the opposite side of the road from the settlement. The unique position of the settlement is regularly highlighted during bigger protests, as the police often use the streets around Jazdow to park their radio cars.

As state power and money coalesce here, it was not surprising that the mayor's vision of 'acceptable' functions claimed to be 'more adequate' for the area included a new Japanese embassy and a shopping mall. This process interestingly coalesces with other well-described gentrification-like cases in the EE region through 'urban renewal' and 'urban revitalisation' practices, with strong state or municipal presence acting as a catalyst for the changes (Drozda, 2017; Jelinek, 2020; Pastak and Kahrik, 2016). Another peculiarity of Warsaw and many other cities in the region is that 'gentrification' processes occur in sync with shifts in property regimes, either on formerly public land or due to ownership changes and restitutions (often filled with irregularities).

Around this time, a group of citizens and Jazdow inhabitants started to organise to defend the area from development, forming an 'Open Jazdow' initiative. The goal of the activists was to preserve the living space for the few remaining inhabitants and to keep the area open for everyone, together with its wild character, without overt landscaping or hierarchical organisation, much in line with the Lefebvrian 'right to the city'. In this manner, the area would continue to serve as a counterpoint to the nearby baroque park with sterile and meticulous up-keep and set visiting hours.

What played into the activists' hands was the silent change that took place at the city hall. The liberal administration that had been in power for twenty years with only a minor break saw its approval rate plummet, especially after a far-reaching scandal connected to the restitution of public spaces and housing in the city. Additionally, within Poland, Warsaw city hall was one of the last strongholds of the once powerful liberal party that had enjoyed only very limited power on a national level, as national politics were dominated by the social-conservative Law and Justice (PiS) Party.

The shaky position that the administration found itself in was echoed in their inclination towards experimentation with policy. As Brenner and Theodore (2002: 28) remind us, cities have become absolutely central for policy experiments and new politico-ideological projects designed to deliver changes that in reality allow 'business as usual' to continue. While the earlier rule of the liberal party could be described as full-throttle neoliberalism, it soon became obvious that the revanchist approach (Smith, 1996) of aggressively uneven and exclusionary urban governing had met its limits. In addition, the GFC[7] slowed down the influx of investment, forcing the city hall to limit its spending. The city hall found the solution to this conundrum in the creative city paradigm, which the city hall hesitantly adopted (see Florida, 2005). As several researchers have shown, after the recession, the desire for relatively cheap 'quick fixes' across cities increased, leading to a boom in officially-sanctioned, so-called tactical urbanism (Mould, 2014) and 'creative' approaches (Pratt and Hutton, 2013).

Fig. 3. The green space of Jazdow. Photograph by the author (2018).

7 Poland weathered the financial crisis rather well, but it could nevertheless be felt due to slowed-down investment.

Based on the communications and the new projects of the city hall, as well as their hiring choices, it became clear that the city realised that to perpetuate the image of Warsaw as a cool and modern global city, they would need a few 'alternative'-looking areas, with people riding bicycles and at least a simulacrum of 'authentic' urban experiences;[8] delivering the urban lifestyles to attract the middle- and upper-class inhabitants following developments in other global cities. For example, Colomb (2012) has analysed policy shifts in urban governance in Berlin, a city that acts as one of the main sources of inspiration in Poland due to perceived geographical and cultural similarities, and even includes a discussion on the city's use of the urban 'voids' in urban development and marketing of space. Similar tendencies of rebalancing neoliberal development have been described in other cities around the world, e.g. in an article on urban development in Moscow by Zupan and Budenbender (2019) as "hipster Stalinism", a way of rejuvenating and greening space physically, without any attempt to address deeper undemocratic rules or strategies inherent to the processes of urban development politics and planning.

To keep up with the "global economy of appearances" (Tsing, 2000), several well-known urban experts and activists were hired by the Warsaw city hall. The newly hired consultants perpetuated the creative language following western European urbanist tendencies (and the new head architect gave a Powerpoint presentation called '*Copenhagenize Warsaw*'), although it is questionable whether they enjoyed any reasonable influence to push through their more radical propositions. Generally, the whole story echoed a familiar scenario of the counter-culture representatives in the urban realm being coopted by capital-backed political representation and thus effectively silenced (Frank, 1997). At the same time, however, it was undeniable that the desired mode of urbanity shifted and provided a degree of fluidity that led to the Jazdow initiative's success.

Tapping into the tendencies in the city council, the protest against demolishing Jazdow immediately took the form of 'community activation'. Besides the more conventional forms of struggle, the activists organised concerts, picnics, workshops, and started a community

8 As suggested by one of my informants, a young female architect.

garden in order to present the space as a valuable gathering spot for the community. At the same time, they employed the language that echoed the 'creative city paradigm' and suggested that Jazdow could be an urban laboratory of a kind. Some of the otherwise rather radical positions remained strategically under-communicated[9] while the activities that fitted well with the city hall agenda were highlighted. The Open Jazdow group also made sure that they had a professional and very neat-looking visual representation of their plans and attempted to include 'respectable' organisations and experts[10] in their propositions.

There is a significant corpus of academic literature that shows how the creative city discourse often serves as a way to introduce and justify gentrification in impoverished areas (McLean, 2017: 41–42; Pratt, 2009; Zukin, 1995). According to this line of reasoning, public activation, artistic interventions or community gardening herald the upcoming 'clean-up' of neighbourhoods that turns them into a desirable destination for capital. The irony in the case of Jazdow is that these dynamics have curiously worked the other way around—as a safeguard against being pulled into a pre-existing government-business district with its sterile, elitist space. The discourse that the activists more or less willingly coopted thus acted as a functioning buffer that succeeded in persuading the city *against* development. It also ensured that the area could be kept messy and open, and created space for more radical experiments that would otherwise never take place.

Warsaw: The Green Capital?

Besides the space activation and creative city discourse, 'gap activism' has acquired a new tool, as Warsaw has seen a sudden and somewhat surprising turn towards environmental issues. The spike was signified by an air pollution crisis that has been alarming for decades, but for various reasons only truly became a household concern during the

9 An example of this was the pilot micro-housing initiative for the homeless or some of the commoning experiments.

10 For example, at the moment some of the houses are occupied by well-established NGOs. One provides support for migrants, another works in education, and there is a group of young architects. At the same time people who presented themselves as experts on urban topics, urban ecology and more have been involved in order to gain a respectable and favourable negotiating ground.

winter of 2016–2017. One environmental activist described his feelings about the sudden buzz around his work by likening his organisation to a rock band that had been confined to their garage for years before getting 'discovered'. Throughout the winters, EU norms on air pollution were exceeded almost every day. In January 2017, for one day, Warsaw even topped the global air pollution statistics. Linking this to the aforementioned attempts to mimic the West, it could be argued that the January smog was a breaking point. Newspapers printed titles suggesting that Warsaw was worse than Beijing, to emphasise a cultural horror about being so far from desired western standards and core.

Amidst barely existent industrial pollution in the capital region, the culprits were found to emanate from two of Poland's greatest loves: cars and coal. Leading up to this period, between 2009 and 2014, approximately 147,000 trees were cut from within Warsaw, often with justification and reference to the 'modernisation' of public space in Warsaw ('Urban heat island' report, 2014). However, with the upheaval over pollution, environmental concerns entered the agenda of both inhabitants and municipal governments. In its attempts to search for new branding ideas, the city hall even entered a competition 'Warsaw—the green capital of Europe' with reference to the large numbers of parks and "lush green space" in the city (Jakubowski, 2015). Even though the green capital slogan was laughed off by many as a poor attempt at greenwashing, it nevertheless remained in the official municipal communications, with the promotional video being played on public transport and regular announcements by the administration about Warsaw as the "green capital in-the-making" (most recently in a tweet by Mayor Trzaskowski in May 2018). Similar to other cases described by the literature (Rademacher, 2008; Szili and Rofe, 2007), environmental issues made a grand entrance in Warsaw and became focal points for manifesting rifts in state policy-making.

Amid the upheaval that the city simply could not ignore, the Jazdow area as well as other gaps in the city started gaining new salience as spaces of urban reproduction, as the sensibilities of the media as well as inhabitants became more attuned to environmental issues and greenery in various shapes and forms. The arguments at hand for a long time amongst activists suddenly gained power and coverage. The activists were now able to discuss how the vacant lots could vitally contribute towards urban living as well as providing a refuge for vagabond species of plants and animals (Harrison and Davies, 2012; Clement, 2011). They discussed how Jazdow and similar areas mitigated urban heat island issues, and even tapped into urban ecology theorisations and presented these as options to be included in the official city policies, e.g. in the ongoing 'Adapcity' project acting as an umbrella institution for research on resilience and climate change in Warsaw. Ironically, the above-mentioned policies that had kept the gaps devoid of social action and allowed for the overgrown greenery to develop contributed to making such spaces even more valuable once urban space was reframed. The pollution crisis revealed the ruderal ecologies of the gaps as vital urban spaces, finally transgressing the dated nature-urbanity duality.

Moreover, the area of Jazdow, as well as many other gaps, underwent a certain metamorphosis of scale. Earlier, Jazdow had been perceived as an isolated issue—today it is viewed as part of a system of Warsovian green corridors. Almost forgotten in the 1990s, Warsaw boasts a system of 'air wedges' filled with unkempt greenery, parks, cemeteries or airports, which have been in place since the war—and which are the reason behind many of the gaps in the city's fabric. As nowadays the corridors are often built over, the topic of their usefulness for air ventilation rose to prominence in relation to the air-pollution crisis. Within a surprisingly short time span, the loss of green wedges was singled out as a symbol of wild capitalism, which should be prevented.[11] Jazdow is located at the tip of one of the corridors, linked to the system of public parks that penetrates the city centre.[12] Due to this positioning, Jazdow gained an unequivocal boost in the way the legitimacy of its existence was perceived by the Warsovians. As earlier rights-to-the-city approaches became intertwined with pressing environmental

11 The existence of the corridors has always been included in urban plans, however these were not binding and officials rarely respected them in their building permit decisions.

12 While the effect of trees and greenery on the urban environment can be scientifically demonstrated, at the time of research nobody in Warsaw seemed able to conclusively comment on whether the green corridors actually work in their main task of circulating fresh air within the city. As one researcher dealing with air pollution told me, the corridors in Warsaw are simply a matter of tradition and "people sort of believe in them." Nevertheless, even without clear evidence, they have become a buzzword, a sort of long-lost panacea to smog invoked by environmentalists and the leftist opposition alike.

grievances, it became difficult to imagine that the city hall would move towards developing the area. Urban greenery in the gaps and elsewhere gained a new political and practical salience that could not be ignored.

***

The literature posits depoliticisation or even a post-political agenda of green urban projects (Swyngedouw, 2018). However, at the time of the Jazdow case, the city of Warsaw was still reluctant to embrace the green sustainability paradigm, and environmental struggles had a unique aura of veiled politics, hiding deeply political agendas behind seemingly apolitical issues. By using gaps as a means in urban struggles, an intriguing circle in Polish activism has been drawn: the current situation echoes the activist scene of the 1980s around the EE region. As the 1980s movements realised, environmental activism is a topic through which it continues to be possible to launch systemic critiques of government and policies, by raising seemingly 'apolitical' grievances. In a deeply divided Polish society, the Jazdow initiative has thus attained a unique position from which it has become possible to critique and highlight other far-reaching topics such as the Polish coal industry, automobile fetishism or climate change, and link them to uneven development in the city and beyond.

Conclusion

Through gaps and the interstitial areas in the city, I have explored how policy decisions in post-socialist and post-crisis Warsaw have yielded contradictory results. At different points, the imaginary of a global city in the neoliberal era and unique property relations have combined with a modernist urban outline that has ironically led to both the appearance and disappearance of urban gaps. While they have caused a large degree of dormancy within such spaces, they have also provided an opening for city residents to 'reclaim' them.

The Jazdow initiative exemplifies one case study of such struggles towards the preservation and rearticulation of urban gaps in Warsaw. By proposing an alternative approach to gaps, opening them up and highlighting the value of urban greenery, the Jazdow community showed that an eventual reconciliation with the goals of municipal policy-makers was possible. Jazdow achieved this by taking advantage of a policy shift at the city hall that embraced a paradigm of creative city and public 'activation'. By making use of the new approach to urbanism, Jazdow activists essentially found a tool that enabled the preservation of accessible and non-commodified areas within the heart of one of the most exclusionary and securitised neighbourhoods. The recent rise of the environmental agenda has further supported the efforts of the initiative and provided the community with a chance to link their struggle to more systemic issues reaching far beyond Warsaw.

It must be noted that until now, Jazdow activists have been rather shy to use the full potential ushered in by the environmental crisis and to truly jump scales in their work. In *Cognitive Cartography in a European Wasteland*, Gille (2000) described the competing forms of invocations of the desired 'Westernness' in a struggle over a trash processor. In my case, the competing imaginaries of urbanity are only implicit, as one side tactically waits and uses the cracks in the other's discourse. It would be easy to criticise the approach for a lack of radicalism—after all, in order to maximise capital extraction in urban areas, 'weird' and alternative spaces in the city are currently highly sought after by policy-makers. However, after decades of tough policing as well as harmful and unfavourable city hall policies, the strategy could also be described as mimicking the power imbalance between the two sides of the struggle. The activists were manoeuvered into carefully testing their field of possibilities before unleashing action and thus far, this strategy has been rather successful. At the same time, in the near future it will be immensely important to develop the range and scale of topics that the initiative touches on, to continue with a simple territorial expansion to other similar areas in the city and the country—before they disappear, as so many already have before them.

References

Anderson, J., 2015. *Understanding Cultural Geography: Places and Traces*. Routledge, Oxford. https://doi.org/10.4324/9781315819945.

Bell, D., and De-Shalit, A., 2014. *The Spirit of Cities: Why the Identity of a City Matters in a Global Age*. Princeton University Press, Princeton, NJ. https://doi.org/10.1515/9781400848263.

Berger, A., 2006. *Drosscape. Wasting Land in Urban America*. Princeton Architectural Press, New York. https://doi.org/10.3368/lj.27.1.154.

Brenner, N., and Theodore, N., 2002. Cities and the Geographies of 'Actually Existing Neoliberalism', in: Brenner, N., and Theodore, N. (Eds.), *Spaces of Neoliberalism: Urban Restructuring in North America and Western Europe*, Blackwell, Oxford, pp. 2–32. https://doi.org/10.1111/1467-8330.00246.

Carlsson, C., and Manning, F., 2010. Nowtopia: strategic exodus? *Antipode*, 42, 924–53. https://doi.org/10.1111/j.1467-8330.2010.00782.x.

Choy, T., 2011. *Ecologies of Comparison: An Ethnography of Endangerment in Hong Kong*. Duke University Press, Durham, NC. https://doi.org/10.2307/j.ctv125jmf0.

Clément, G., 2011. In Praise of Vagabonds. *Qui Parle*, 19(2), 275–97. https://doi.org/10.5250/quiparle.19.2.0275.

De Sola Morales, I., 1995. Terrain Vague, in: Davidson, C., (Ed.). *AnyPlace*, MIT Press, London, pp. 118–23.

Colomb, C., 2012. *Staging the New Berlin: Place Marketing and the Politics of Urban Reinvention Post-1989*. Routledge, London and New York. https://doi.org/10.4324/9780203137543.

Derrida, J., 2006 (1993). *Specters of Marx. The State of the Debt, the Work of Mourning and the New International*. Routledge, London and New York. https://doi.org/10.5040/9781350250307-0200.

Drozda, Ł., 2017. *Uszlachetniając przestrzeń: jak działa gentrykacja i jak się ją mierzy*. Instytut Wydawniczy Książka i Prasa, Warsaw.

Dubeaux, S., and Sabot, E. C., 2018. Maximizing the potential of vacant spaces within shrinking cities, a German approach. Cities, 75, 6–11 https://doi.org/10.1016/j.cities.2017.06.015.

Edensor, T., 2005. *Industrial Ruins. Spaces, Aesthetics and Materiality*. Berg, Oxford and New York. https://doi.org/10.5040/9781474214940.

Fehervary, K., 2002. American Kitchens, Luxury Bathrooms, and the Search for a 'Normal" Life in Postsocialist Hungary, *Ethnos*, 67(3), 369–400. https://doi.org/10.1080/0014184022000031211.

Florida, R. L., 2005. *Cities and the Creative Class*. Routledge, New York. https://doi.org/10.4324/9780203997673.

Frank, T., 1997. *The Conquest of Cool: Business Culture, Counterculture and the Rise of Hip Consumerism*. University of Chicago Press, Chicago, IL. https://doi.org/10.7208/chicago/9780226924632.001.0001.

Gandy, M., 2009. Interstitial landscapes: Reflections of a Berlin corner, in: Keil, R. (Ed.), *Urban Constellations*. Jovis, Berlin, pp. 149–52.

Gille, Z., 2000. Cognitive Cartography in a European Wasteland: Multinational Capital and Greens Vie for Village Allegiance, in: Burawoy, M. et al. (Eds.), *Global Ethnography*. University of California Press, Berkeley, CA, 240–67.

Hann, C., 2005. Postsocialist Societies, in: Carrier, J. G. (Ed.), *A Handbook of Economic Anthropology*. Edward Elgar Publishing, Cheltenham, pp. 547–58. https://doi.org/10.4337/9781845423469.00053.

Harrison, C., and Davies, G., 2002. Conserving biodiversity that matters: practicioners' perspectives on brownfield development and urban nature conservation in London. *Journal of Environmental. Management*, 65, 95–108. https://doi.org/10.1006/jema.2002.0539.

Harvey, D., 2012. *Rebel Cities: From the Right to the City to the Urban Revolution*. Verso, London.

Heynen, N., 2014. Urban Political Ecology I: The Urban Century. *Progress in Human Geography*, 38, 598–604. https://doi.org/10.1177/0309132513500443.

Hirt, S., 2012. *Iron Curtains. Gates, Suburbs and Privatization of Space in the Post-Socialist City*. Wiley-Blackwell, Chichester. https://doi.org/10.1002/9781118295922.

Huyssen, A., 1997. *The Void of Berlin*. University of Chicago Press, Chicago, IL.

Jansen, S., 2015. *Yearnings in the Meantime: Normal Lives and the State in a Sarajevo Apartment Complex (Dislocations)*. Berghahn Books, New York, Oxford. https://doi.org/10.2307/j.ctt9qcxhw.

Jakubowski, T., 2015. Warszawa zostanie Zieloną Stolicą Europy? Jest już na półmetku. *Wawalove*, https://wawalove.wp.pl/warszawa-zostanie-zielona-stolica-europy-jest-juz-na-polmetku-6178738287978113a.

Jelinek, C., 2020. Turning a "Socialist" Policy into a "Capitalist" One: Urban Rehabilitation in Hungary during the Long Transformation of 1989, *Journal of Urban History*, 47/3, 511–25. https://doi.org/10.1177/0096144220908880.

Kusiak, J., 2012. The Cunning of Chaos and Its Orders, in: Grubbauer, M., and Kusiak, J. (Eds.), *Chasing Warsaw. Socio-Material Dynamics of Urban Change since 1990*, Campus Verlag, Frankfurt, New York, pp. 291–320.

Landry, C., 2000. *The Creative City: a Toolkit for Urban Innovators*. Earthscan, London.

McLean, H., 2016. Hos in the garden: Staging and resisting neoliberal creativity. *Environment and Planning D: Society and Space*, 35 (1), 38–56. https://doi.org/10.1177/0263775816654915.

Mould, O., 2014. Tactical Urbanism: The New Vernacular of the Creative City. *Geography Compass*, 8, 529–39 https://doi.org/10.1111/gec3.12146.

Murawski, M. 2018. Marxist morphologies. *Focaal* 2018, 82, 16–34. https://doi.org/10.3167/fcl.2018.820102.

Novy, J., and Colomb, C., 2013. Struggling for the Right to the (Creative) City in Berlin and Hamburg: New Urban Social Movements, New 'Spaces of Hope'? *International Journal of Urban and Regional Research*, 37, 1816–838. https://doi.org/10.1111/j.1468-2427.2012.01115.x.

Pastak, I., and Kahrik, A., 2016. The Impacts of Culture-led Flagship Projects on Local Communities in the Context of Post-socialist Tallinn. *Sociologický Časopis / Czech Sociological Review*, 52(6), 963–90. https://doi.org/10.13060/00380288.2016.52.6.292.

Phelps, N. A., and Silva, C., 2018. Mind the gaps! A research agenda for urban interstices. *Urban Studies*, 55(6), 1203–22. https://doi.org/10.1177/0042098017732714.

Pobłocki, K., 2012. Class, Space and the Geography of Poland's Champagne (Post)Socialism, in: Grubbauer, M., and Kusiak, J. (Eds.), *Chasing Warsaw. Socio-Material Dynamics of Urban Change since 1990*, Campus Verlag, Frankfurt and New York, pp. 291–320.

Pratt, A. C., 2009. Urban regeneration: from the arts 'feel good' factor to the cultural economy. A case study of Hoxton, London. *Urban Studies*, 46(5–6), 1041–61. https://doi.org/10.1177/0042098009103854.

Pratt, A., and Hutton, T., 2013. Reconceptualising the relationship between the creative economy and the city: learning from the financial crisis. *Cities*, 33, 86–95. https://doi.org/10.1016/j.cities.2012.05.008.

Rademacher, A., 2008. Fluid City, Solid State: Urban Environmental Territory in a State of Emergency, Kathmandu. *City & Society*, 20 (1), 105–29. https://doi.org/10.1111/j.1548-744x.2008.00008.x.

Saksouk-Sasso, A., 2013. Contesting National Authority in the Construction of Public Space: the making of communal spaces in Beirut. Paper presented at the annual meeting for the Open Air Space Gathering, Beirut, Lebanon, December 2013.

Schindler, S., 2014. The making of "world-class" Delhi: Relations between street hawkers and the new middle class, *Antipode*, 46(2), 557–73. https://doi.org/10.1111/anti.12054.

Sieverts, T., 2001. *Zwischenstadt. Zwischen Ort und Welt, Raum und Zeit, Stadt und Land*. Birkhäuser, Basel, Boston, MA and Berlin.

Smith, N., 1996. *The New Urban Frontier: Gentrification and the Revanchist City*. Routledge, New York. https://doi.org/10.4324/9780203975640.

Stavrides, S., 2014. Open space appropriations and the potentialities of a 'city of thresholds', in: Mariani, M., and Barron, P. (Eds.), *Terrain Vague: On the Edge of the Pale*. Routledge, London, pp. 48–61.

Stenning, A., and Hörschelmann K., 2008. History, geography and difference in the post-socialist world: Or, do we still need post-socialism? *Antipode*, 40 (2), 312–35. https://doi.org/10.1111/j.1467-8330.2008.00593.x.

Swyngedouw, E., 2018. *Promises of the Political: Insurgent Cities in a Post-Political Environment*. MIT Press, Cambridge, MA. https://doi.org/10.7551/mitpress/10668.001.0001.

Szili G., Rofe W. M., 2007. Greening Port Misery: Marketing the Green Face of Waterfront Redevelopment in Port Adelaide, South Australia. *Urban Policy and Research*, 25 (3), 363–84. https://doi.org/10.1080/08111140701540695.

Stoetzer, B., 2018. Ruderal Ecologies: Rethinking Nature, Migration, and the Urban Landscape in Berlin. *Cultural Anthropology*, 33(2), 295–323. https://doi.org/10.14506/ca33.2.09.

Tonnelat, S., 2008. 'Out of frame:' The (in)visible life of urban interstice. *Ethnography*, 9 (3), 291–324. https://doi.org/10.1177/1466138108094973.

Tsing, A. L., 2000. Inside the economy of appearances. *Public Culture*, 12 (1), 115–44. https://doi.org/10.1215/08992363-12-1-115.

Urban Heat Island Report (Miejska wyspa ciepla) (2014). Polish Academy of Science and Institute of Geography and Spatial Planning S. Leszvzyckiego, Wydawnictwo Akademickie SEDNO, Warszawa, www.ecin.org.pl.

Verdery, K., 1998. Property and Power in Transylvania's Decollectivization, in: Hann, C. M., (Ed.), *Property Relations: Renewing the Anthropological Tradition*. Cambridge University Press, Cambridge.

Vivant, E., 2013. Creatives in the city: Urban contradictions of the creative city. *City Culture and Society*, 4, 57–63. https://doi.org/10.1016/j.ccs.2013.02.003.

Zydek, S., 2014. 'Warsaw' in Vacant Central Europe: Mapping and recycling empty urban properties, in: Polyak, L., (Ed.), *Vacant Central Europe: Mapping and Recycling Empty Urban Properties*. Kek, Budapest. https://issuu.com/kekfoundation/docs/vacant_central_europe.

Zukin, S., 1995. *The Cultures of Cities*. Blackwell, Cambridge, MA.

Zupan, D., and Büdenbender, M., 2019. Moscow urban development: neoliberal urbanism and green infrastructures, in: Tuvikene, T., Sgibnev, W., & Neugebauer, C. S. (Eds.). *Post-Socialist Urban Infrastructures* (OPEN ACCESS) (1st ed.) Routledge, Oxford, pp. 125–41. https://doi.org/10.4324/9781351190350-8.

PART II

5. Far-right Grassroots Environmental Activism in Poland and the Blurry Lines of 'Acceptable' Environmentalisms

Balsa Lubarda

More than a decade ago, the rise and subsequent mainstreaming of right-wing populism and far-right politics seemed unimaginable to many. In the countries of Central and Eastern Europe, articulating nationalism as the 'reclaiming of politics' and the panacea to transitional hardships has brought populist (and) far-right parties to major electoral successes. This has allowed a number of far-right actors (not necessarily parties) to engage in various debates, including topics related to the natural environment. This chapter seeks to determine how far-right groups in Poland, *Ecolektyw* (formerly *Greenline Front Polska*) and *Puszczyk-Naturokultura Polska* (also associated with *Praca Polska*) convey their ideological positions concerning the natural environment, as well as how this ideological content becomes embedded in and amended through activism.

The political far right in Poland has evolved with, but also in opposition to, the populist radical right. Following *Fidesz*, their successful role model in Hungary, the Polish Law and Justice Party, *Prawo i Sprawiedliwość* (PiS), used the Manichean, binary logic of the 'good people' and a 'corrupt elite' to sweep away most of their political opponents. Ever since 2005, when PiS first came to power under the banner of a "moral revolution" (Harper, 2018: 29), the party has purposefully tried both to entrench its political power and monopolise

 https://doi.org/10.11647/OBP.0244.05

historical narratives, asserting its status as the defender of traditional values. In spite of PiS's alleged intention to have "only the wall" to their right (Harper, 2018: 59), the diversified far-right[1] landscape in Poland has served predominantly as the opposition to the ruling party, particularly since 2015 when PiS re-entered government. As the far-right party opposition to PiS is relatively formally weak (*Ruch Narodowy* with six members of *Sejm* in the Lower House of Parliament integrated in the *Konfederacja* coalition), it devotes its attention to topics that are electorally lucrative, such as migration, EU regulations and questions of collective identity.

However, this is not the case with far-right movements, which regardless of their limited financial resources, have greater space for manoeuvring and engaging with various topics, such as the environment. This not only highlights particular social imaginaries (Castoriadis, 1975) but also bears a substantive and ideological morphology (Freeden, 1996), distinguishable from right-populist accounts on the environment. For instance, the right-wing populist emphasis on protecting 'the people' often used as a justification for anti-environmental stances and policies, is notably different from the far right's endorsement of environmental protection through the intricate relationship of the nation and the land (see Forchtner and Kølvraa, 2015). Nevertheless, both the right-wing populist party in power (*PiS*) and far-right parties in opposition (*Ruch Narodowy—Konfederacja*) have been recognised either as climate sceptics or as outright deniers of anthropogenic climate change (see Lockwood, 2018: 715; Żuk and Szulecki, 2020). This attitude is in line with post-materialist renderings of environmentalism as being an (unwanted) offspring of democratic transformation (Inglehart, 1971).

In contrast, there are far-right movements in Poland specifically focused on environmental activism. The far-right's propensity for grassroots organising (for examples, see Castelli, Gattinara and Froio,

1 This can partly be attributed to problematic definitional properties of the far right, which is itself a contested term, overhauled by ambiguities and malleable features. I will here employ the term 'far right' instead of some other catchall terms used by scholars, such as the radical right, extreme right, hard right, right-wing populism, radical nationalism, neofascism, etc. Far right is the broadest of these terms, as it comprises the extreme right (neo-Nazis and other groups openly aiming to overthrow the democratic system) and the radical right (still operating within the boundaries of liberal democratic political systems).

2018), particularly the ways in which this grassroots engagement unfolds in eastern Europe (see Mikecz, 2015), renders these groups suitable for environmental activism. In eastern Europe, to an extent incomparable with other regions, the far right has managed to transform into 'movement parties', building on the potential of their local structures. Nevertheless, this potential is hindered by the ideological differences that exist between the far right and most green activists. Grassroots environmentalism is commonly associated with contentious politics and environmental justice, providing an apt opportunity for constructing alternative visions. As such, grassroots and bottom-up organising is often viewed as a blueprint for a 'progressive' and 'emancipatory' change (see Borras, 2019).

To understand the implications of far-right local environmental activism for its influence on collective identity and action, this chapter has the following aims. First, it explores the forms of far-right incursions into environmental thought, by looking at the content or the ideological morphology of 'Far-Right Ecologism' (FRE; Lubarda, 2020). Second, the chapter explores how FRE has been adjusted to local contexts, or how some of its elements have been assuaged and normalised by local networks of environmentalists in order to increase their support base. To do so, this chapter will first outline the ideological morphology of FRE, which attempts to situate the far-right ideological position in relation to the natural environment. This will allow for a zooming-in on the respective ecologisms of *Ecolektyw* and *Puszczyk*, before analysing how these groups interact with other grassroots organisations.

The Political Far Right and Environmentalism: From Climate Denialism to Far-Right Ecologism?

The nexus of the political far right and environmental issues has only recently regained substantial attention from scholarship interested in the development of the formal far right (for examples, see Forchtner, 2019a and 2019b; Voss, 2014). However, the vast majority of scholarly works dealing with the contemporary far right and the environment focus most on far-right climate denialism (see Krange, Kaltenborn, and Hultman, 2018; Anshelm and Hultman, 2014).

The potential reason for the lack of sustained engagement with this topic is the somewhat unimaginable association of the far right with environmental activism, which has roots that predate traditional eco-fascism (see Bruggemeier et al., 2005). In fact, the content of today's far-right ideological stance on this issue stems from nineteenth-century writings on the environment, predominantly the *völkisch* and ethnonationalist traditions that contributed to the coinage of the eco-fascist 'Blood and Soil' concept (Bramwell, 1985). The 'patriotic duty' to protect local and national environments may easily bear resemblance to the convergence of naturalism, organicism, and authoritarianism (Olsen, 1999), which are vivid in fascist ideals of national rebirth (Griffin, 1991). Moreover, organic farming has had its tributaries among the fascists and Nazis across Europe (for Germany, see Bramwell, 1985, for the UK, see Coupland, 2017, or for Hungary, Lubarda, 2020). Thus, aligning nativist, anti-immigrant sentiments with eugenic policies has certainly not been immanent only to the extreme right: occasionally, such links can be found amongst deep ecologists as well (see Staudenmeier, 2011).

Although eco-fascism has long been considered the most suitable term to describe cases of pro-green, far-right actors, it falls short of capturing the complex ideological content of the far right regarding the environment, which is derived from a variety of right-wing ideologies (such as conservatism, nationalism, and right-wing populism). For instance, the profound and abstract connection of nationalism and space (Smith, 1991) has led to an appreciation of the local environment, through conservative tropes of responsibility for and love of the home, termed *oikophilia* (Scruton, 2012). Moreover, the landscape has provided ample material for cultural mediation and reproduction in nationalism, ultimately leading to depictions of authentic, national environments as self-sustaining ecosystems (Sorlin, 1999; Forchtner and Kolvraa, 2015). The endorsement of localism, as opposed to abstract 'globalist' positions, have prompted some authors to imagine ecological forms of nationalism as potentially progressive and desirable (Gare, 1995; Dawson, 1996). Likewise, Barcena, Ibarra, and Zubiaga (1997: 302) argued that nationalism and environmentalism overlap in their defence of the local and particular.

However, the inwards-oriented mysticism associated with these outlooks can also lead to nativist sentiments epitomised in the perceived threat of foreign races and cultures under the banner of "preserving the ecosystem" from "foreign or invasive" species (Olwig, 2003: 61). This naturalist logic of purity lends itself to existing criticisms of human migration. The desire for order and stability, for clear (even if symbolic) boundaries, has been at the core of the vision of "polluting outsiders" (Lubarda, 2017), pointing to how environmental nativism may unfold (see Forchtner, 2019a).

From the anthropocentric 'oikophilia' to the populist struggle against "environmental elites" (Szasz, 1996), the far right's ideological morphology tailored to the environment requires a more comprehensive analytical framework. Simply calling these incursions 'eco-fascist' is insufficient and fails to account for the multifarious (and yet distinctive) ideological views of how perceptions of the environment inform the far-right worldview. The notion of 'Far-Right Ecologism' (FRE) incorporates the broader right-wing spectrum in its ideological morphology (see Lubarda, 2020). The core concepts of this ideology revolve around binary, Manichean distinctions between 'good' and 'evil' (good nationalists vs. 'evil' capitalists/liberals), naturalism (viewing nature as a blueprint for social order), and organicism (the notion of nation, culture, and nature in a holistic union as a single organism). These elements are indispensable, and thus present in all possible variations and instances of FRE. Relevant adjacent and peripheral concepts to this conceptual core are nostalgia (for example, calls for a return or a 'rebirth' of the imagined ecological polity of the past by fostering 'traditional' practices such as family farming), autarky, mysticism and spirituality (from the polytheism or paganism of the eco-fascists, to Christian ecologists' view of nature as God's gift), and authority (survivalism or decentralisation through a 'family of families'). These elements may constitute FRE as an ideal type, a heuristic device that enables the identification of distinctive features of the far-right groups under scrutiny below.

The data used in this chapter originates from fieldwork conducted in Poland between November 2018 and September 2019. The corpus consists of participant observation notes and eight qualitative, semi-structured interviews with representatives of far-right organisations (*Ecolektyw*, *Puszczyk*, *Praca Polska*), as well as social media posts and media items posted on Polish online portals related to the environmental activism of these organisations, or their cooperation therein. As FRE is envisaged as a loose analytical framework that helps the coding process, the analysis will also reflect on the interactions of activists with the topics they are focused on, as well as other relevant actors in the process.

Ecolektyw and *Puszczyk*: The 'Other' Type of Grassroots Organising

Ecolektyw is an informal, self-proclaimed eco-nationalist organisation operating in Poland. It originated in 2016, from *Greenline Front Polska*, the Polish chapter of an international eco-fascist movement formed by a group of Ukrainian nationalists. Greenline Front was established as a loose, leaderless, and 'memberless' organisation, founded by "national-conscious people, who reject the system of modern anthropocentric values and its procreations, such as capitalism and monotheistic religions" (*Ecolektyw*, Facebook, 15 May 2018). In addition to these spiritual underpinnings, the organisation uses the slogan 'blood-soil-nature', coupled with esoteric mysticism of *völkisch* nationalism. The logo of *Ecolektyw* is emblematic of FRE's appropriation of the runic alphabet, characteristic of identitarian and neo-Nazi movements. The 'Life Rune' (Algis or Elhaz) is the central part of the logo, and originates from early Germanic and Nordic alphabets, which were later associated with the Nazis (in particular, the SS's *Lebensborn* project), but has also appeared in diverse neo-pagan and occult movements around the world (see Dahmer, 2019).

The materials posted on the 'official' blog of the Greenline Front caution against "harmful racial policies" that posit a danger to the imagined equilibrium of nature and human beings (Greenline Front Blog, 2019). The blog also comprises quotes from deep-ecologists such as Pentti Linkola, denoting democracy as the "religion of death" (Linkola, 2019), but also some less controversial conceptual connections such as that of Leonardo da Vinci's ethical vegetarianism. In the self-acclaimed "revolt against the modern world", Greenline Front evokes the tenets of the *völkisch lebensreform* (life-reform) movement, such as a back-to-nature lifestyle, organic production, but also underlying asketism and Manicheanism. Within this cosmology, the fundamental aim of Greenline Front (and later, *Ecolektyw*) was to combat the capitalist vision of a "stateless person who wants us all reduced to consuming larvae"

(Greenline Front, 2018). Although the Greenline Front had relative success in its first year, the movement has slowly died out since, with the majority of its influential informal leaders hampered by a combination of personal problems, the pressure of authorities and the media.

After a brief period of inactivity, in May 2018 Greenline Front Polska changed its name to *Ecolektyw* (ecological collective), which in the words of one of its founders, aimed to "establish a new trajectory and form of organising as compared to strictly nationalist Greenline Front tenets" (Interview with Rex, 15 June 2019). This alleged move away from established nationalist, far-right tenets was primarily induced by the lack of interest of the Polish Autonomous Nationalists (who constituted the majority of this movement), as well as organisational obstacles caused by the Greenline Front's previous ideological standpoints.

Unlike *Ecolektyw*'s (openly) eco-fascist foundations, *Praca Polska* (Poland Works) and *Puszczyk* (in English, tawny owl)—*Naturokultura Polska* are the offspring of attempts to reconceptualise the connection between Polish nationalism and ecological thought.[2] The former was established in 2018, with an aim to deal with the issues of social justice, 'blue collar' and female workers. This essentially 'left' dimension of Polish nationalism (also present in other contemporary eastern European far-right groups) has a notably different image from previous militant, falanga-style marches in black or green shirts (ONR and Autonomous Nationalists as cases in point). It addresses issues associated with neoliberal capitalism as fundamentally problematic and aims to contextualise the multi-faceted nature of problems instead of resorting merely to the revival of a long-lost and supposedly acclaimed past. *Puszczyk* was an environment-focused offspring of cooperation between female members of *Praca Polska* and ONR, aiming to bring together nationalists and environmentalists as an educational link on the imaginary modernist border between "nature and culture" (Puszczyk, 2020). The organisation did not have its formal membership, as most of its activities were organised by the female co-leaders. Since its followers were also a part of other far-right organisations, its relevance

2 This nexus has been recognised by other far-right organisations, such as the All-Polish Youth (*Młodzież Wszechpolska*), which proclaimed the pioneer of nature conservationism in Poland, Jan-Gwalbert Pawlikowski (1891–1962), as their 'patron' for the year 2019.

should be evaluated through its personal networks rather than the link itself. Its eclectic nature brought *Puszczyk* severe organisational hardships, and its webpage is now almost defunct or promoting the activities of *Praca Polska*. With these continuous rearrangements in mind, the environmental activism of these organisations rests on the continued articulation of their ideological tenets. Thus, the chapter will now explore the ideological attempts of the far right to 'reclaim' ecologism.

Environmental Manicheanism: *Ecolektyw* and *Puszczyk*'s FRE

Within the Manichean, binary vision of FRE, the ideal proponent of such an ideology claims to distinguish acceptable practices from unacceptable practices, friends from foes, 'us' from 'others'. Hence, *Ecolektyw* attempts to represent environmental issues that are distinct to Poland. Jettisoning the consumerist mindset, partly based on neo-Malthusian critiques of overproduction and overpopulation, the representatives of movements under scrutiny jointly disdain "pop-environmentalism: conducting activities with a solely promotional, PR-purpose" (Interview with Maria, *Puszczyk and Praca Polska*, 7 September 2019, and Stella, *Ecolektyw*, 2018). To them, the popularisation of environmental concern is a by-product of the "world of excess", as Maria frames it, the condition of today's times incurred by the "consumerist logic of neo-liberal capitalism disparaging natural, ecological communities: the nations". In a similar vein, environmental organisations with global outreach, such as Greenpeace and WWF, are referred to as "money-making corporates" (Kamil, 2019), lacking genuine interest in local struggles. 'Struggle' is a deliberate word-choice here and a point of departure from 'left' grassroots organisations, which becomes paramount through the neo-Spenglerian and somewhat mystic outlook on Manichaeism:

> We declare war on the modern world, and on the laws it tries to impose on us, we will consistently do everything to win this war and help people wake up from a deadly dream.
>
> Let the storm begin, which will bring us the desired victory! Let it bring us the death of the modern world! (*Ecolektyw*'s Facebook page, 2018)

Within this Manichean imaginary, far-right ecologists point to several actors responsible for environmental destruction. Other than the usual culprits (e.g. 'liberals' and 'leftists'), far-right ecologist movements, somewhat unexpectedly, blame the leniency of the Catholic Church for the devastation of the natural environment in Poland:

> There is an ideological problem with Polish Catholicism and the environment—the sheer anthropocentrism, with everything being subordinated to humans including animals [...] so in small cities, where the influence of churches is very big and possibly harmful for these places [...], these human-centred teachings of the church can be problematic for protecting the environment at the moment at the local level (Agata, 2019)

Apart from the issue with the anthropocentric nature of the Church's teachings, such stances are congruent with the alt-right and national anarchist discontent with the hierarchy of clerics. However, the criticisms of *Ecolektyw, Praca Polska,* and *Puszczyk* also point towards their own ranks. Therefore, nationalists (this term is used in self-ascriptions) are perceived as "lethargic", "uninterested", and "wrongly adamant":

> A lot of nationalists still don't realise the importance of this subject. Nationalists associate ecology with LGBT and similar types of stuff because of the activism of these green NGOs and parties. So, they don't want to engage with the topic, which is wrong [...] in nationalist organisations, people who try to talk about the environment, global warming and similar types of topics are laughed at (Małgorzata, 2019).

Another part of these criticisms is aimed at the so-called "salon nationalists" (Zsofia, 2019), quasi-elitist representatives of political parties who disregard the relevance of environmental issues, allegedly giving the topic away to leftists and liberals. Logically, most of these criticisms point to the ruling, right-wing populist *PiS*, but also the far-right parties in opposition, such as *Ruch Narodowy* and partly the All-Polish Youth (*Młodzież Wszechpolska*), who openly dispute scientific findings regarding climate change, and endorse hunting.

So, what exactly is at the core of this applied Far-Right Ecologism? Admittedly, the young leadership of *Ecolektyw* departs from the mysticism of deep ecology (most notably Linkola), identified as problematic by the social ecology scholarship (Biehl, 1989). Apart from the critique of the Catholic Church, the undeniable spiritualism of these movements is based on the appreciation of the esoteric and authentic values of the natural environment. This appreciation is derived from an assumed place-based identity associated with an imagined, national community. However, this appreciation of the local serves as the foundation of grassroots activism. The following passage from Rex, one of the leaders of *Ecolektyw*, explains how this place-based identity melds his engagement with these topics. It also addresses why *Ecolektyw* insists on activities entirely conducted and led by activists from the area immediately affected by a particular environmental issue. In his attempt to disassociate himself from Hitler, Rex nevertheless embraces the *völkisch*, ethnonationalist teachings appropriated by the Nazi party. This explains the 'blood-and-soil', naturalist-organicist imaginary that informs this type of activism, simultaneously disassociating itself with the Nazi regime.

> I appreciate the *Völkisch* tradition not because of Hitler—I did not live in Hitler's time, and Hitler does not live in my time. No, I like the *Völkisch* tradition because I believe in the connection between the nation and the land. Not in the esoteric and magical, mystical sense, but this I derive from my own childhood. I was born in a small city, where there were lots of forests and lakes. My family owned land, they were farmers, and apart from living with cows, I realised how dangerous and severely polluted local rivers became because of profit-seeking individuals and companies, endangering not only our farm but the very idyllic landscape of my childhood. (Rex, *Ecolektyw*, 2019)

This ideological resentment against extractivism and capitalism based on personal experience also sets the ground for topical, issue-based positions of FRE. Unlike the majority of far-right organisations in Poland (including the far-right in western Europe), my respondents are not climate denialists. They tend to be generally supportive of nuclear power as an alternative to coal, which is the primary source of power energy produced in Poland and is considered "black gold" (*Czarne złoto*), typically an important element of national pride (Bridge and Kuchler, 2018: 136).

FRE in Practice: Place-Based Identity as the Common Denominator of the Far-Right and Environmentalists

The use of war-like metaphors paired with mysticism immanent to "right-wing hippies" (Staudenmeier, 1995), such as a spiritual ritual of lighting fire with the symbols of the runic alphabet (as presented on *Ecolektyw*'s Instagram page, 14 February 2020) did not gain substantial attention from the far-right public. After failing to obtain consistent support from the far-right supporters, the leaders of *Ecolektyw* recently indicated an intention to pivot their strategies and collaborate with non-nationalist organisations and movements through an endorsement of 'nationalist' localism.

> I want to make more regional activities. I want to change Szczecin. I feel Szczecin is my second home, so it is important to me that I am not only a Polish man, I am not only a white man, I am not only a European, but [...] I am a person that bears the original culture from the region, the history of Gryf, the specificities of our regional identity [...] I don't want *Ecolektyw* to be all about nationalism, I want the left and the apolitical to be with us. *Ecolektyw* should be more like a hashtag, an idea that brings people together. If some people go to the forest and clean it, or go hiking, take a selfie and put a hashtag #ecolektyw, inviting other people to go, I would be happy. (Rex, *Ecolektyw*, 2019)

In their discursive transition to 'thoughtfulness' instead of 'emotion', as proclaimed on their Facebook page, the 'rejuvenated' *Ecolektyw* has attempted to mediate the Manichean naturalism and organicism with topical, small-scale initiatives in which the 'political' background of their announcers become obfuscated and virtually irrelevant. This image was also bolstered by ecological workshops for youth that are allegedly today the long-term strategy of *Puszczyk's* leaders.

However, the most important step towards broader popularity for these movements was to disassociate (or at least, to make less apparent) the links between their activism and far-right views. While this was not an immense problem for *Puszczyk* or *Praca Polska*, two relatively new organisations, casting away the 'fascist' label for *Ecolektyw* proved laborious. Stranded with the 'eco-fascist' branding as a result of its former association with Greenline Front, the still (relatively) young leadership of *Ecolektyw* decided to rebrand the organisation by avoiding explicit reference to nationalism. This, however, did not influence the types of activities *Ecolektyw* organised and participated in, as the leaders of these organisations continued to partake in small-scale activities to do with local ownership and autarky as individuals (e.g. protesting illegal trash heaps or collecting waste). As they indicated, first-hand experience with air pollution or wastewater and the effort needed to address these issues prompts patriotic feelings of care for one's immediate surroundings. Within this logic, nationalists are more invested in grassroots activism through frames of responsibility, and "those who are not Poles, not even locals, will never feel the space as theirs" (interview with Agata, 2019). This symbolic marker between 'us/good' and 'them/evil' serves as a catalyst, the drive behind ecological actions. These point to how localised affections can be articulated through nationalism to determine those who 'belong' to particular environments.

In strengthening this local-national imaginary, connection with rural communities has proved to be somewhat problematic. This became clear through the issue of animal welfare: the mostly urban membership of *Ecolektyw* and *Puszczyk* attempted to convince rural communities to exclude their plots of land from 'hunting-friendly' zones. Since the members of these communities generally endorsed hunting as an indispensable part of land management, this caused numerous problems for urban-based, far-right organisations. However, this did not prevent *Puszczyk*'s leadership from establishing partnerships with other, non-nationalist, urban grassroots organisations. For instance, *Psia Ekipa* (Dogs' Team, Together for Animals), a local organisation from Bielsko-Biała (in southern Poland), gathered signatures for the construction of animal shelters and helped animals find new homes. *Puszczyk* members also cooperated on the same topic with *Mysikrólik*, a rehabilitation centre for wild animals, led by a group of veterinary physicians and ornithologists, or urban initiatives such as *Bielszczanie dla Drzew* (Bialans for trees), a bottom-up, citizen-led initiative against the shrinking of green areas in the city. This is similar to the initiative in Białystok, *Miasto Mieszkańców* (City of Residents), led by Bogusław Koniuch, a former leader of the now-defunct far-right *Narodowe Odrozenie Polski* (National Rebirth of Poland, NOP), currently a member of the Białystok city council. Koniuch also actively collaborates with

Ecolektyw, and has a proven record of environmental activism, although mostly at the local level of Białystok, including the protection of forests in the wider area of the city (such as the 'ecological picnic in defence of the Turczyński Forest', Koniuch, 2020), including anti-smog protests and moves to preserve urban greenery.

Once again, none of these initiatives or organisations mentioned above originally had nationalist credentials, and most of their members are in fact supporters of opposition parties with liberal or left-leaning ideological grounds. Although some of them were unaware of the ideological background of *Ecolektyw*, *Praca Polska* or *Puszczyk*, and their informal leaders, most of the activists showed indifference towards the worldviews of their collaborators once they were made aware. The involvement of far-right individuals has not significantly changed the ideological profile of local movements, as they continue to remain loose and topic-based initiatives. Thus, much as the examples mentioned in the last paragraph may seem atomised or disjointed, they point to fissures and 'blurry lines' within a seemingly ideologically homogenous environmental movement in Poland, bringing eclectic and mutually distant worldviews on board.

Valuing place-based identities is not a prerogative of the Left, particularly not in eastern European contexts in which nationalism has generally been an important factor in melding agency. Subscribing to far-right ideology does not prevent one from caring for the (local) environment, quite the contrary. Constitutive concepts of FRE, such as nostalgia, are present in other, 'non-right' ecologisms, which enable (former) far-right members to jump on the bandwagon of environmental issues, particularly if they are of a more 'local' nature, or can be presented as a form of empowering self-organisation against environmentally-harmful investments and decisions (for similar examples, see Snajdr, 2008 on the case of *ochranarstvo* in Slovakia).

Although there is tacit support between members, there has been no official cooperation between *Ecolektyw*, *Praca Polska*, and *Puszczyk*. This is partly due to the nature of these organisations, which resemble a rather loose network of nationalists with dedicated informal leaders sharing interests in environmental issues. Members of these groups share the mobilisation networks of Autonomous Nationalists, but also of ONR and formerly NOP: they acquire their ideological positions, iconography, and eventual activism from established far-right networks. In fact, their diffuse and decentralised nature, paired with the transition and collaboration of a range of environmentalist organisations, has severely impacted these far-right collectives. As of 2020, only *Praca Polska* continues to be active, with *Ecolektyw* and *Puszczyk* temporarily defunct. According to the leaders of these groups, such hardships were not caused by their collaboration with environmentalist organisations, but rather with the lack of motivation amongst their wider membership. It is the leaders of these groups, who continue to engage in environmental activism beyond their far-right organisations, that point to how FRE may become integrated and even represented within a broader environmental movement that attempts to resist detrimental environmental practices and policies.

Conclusion

This chapter has explored how an appreciation of the local environment paired with the intention to radically change the current *modus vivendi* of the Anthropocene have enabled the far-right to develop and offer an authentic ideological account of Far-Right Ecologism, reinterpreting and moving beyond eco-fascism. One should not query the sincerity of these outlooks or the discontent of their young leaders (even if attributed to a rebellious, youthful mentality) with the local and national authorities' approach to the environment. This is particularly important in the volatile and competitive far-right landscape, as Poland is not the only eastern European case with competing far-right parties and movements. The competitiveness and continuous struggle for membership and 'survival' in these circumstances drives the propensity of smaller far-right movements towards direct action and grassroots organising. Through the use of an 'oppositional framing', movements such as *Ecolektyw* appropriate Manichean binaries that distinguish between good and evil to communicate urgency and call for immediate action. Moreover, elements of FRE articulated through the outlooks of my respondents, such as nostalgia, organicist holism, and even mysticism associated with nature, are easily tempered under the label of 'environmentalism'.

Since the far right does not exist in a self-sufficient bubble, the movements under scrutiny concentrate on the arenas and interactions that enable them to generate and sustain various forms of resistance. In eastern Europe, far-right groups engaging in grassroots activism with various purposes (such as *Praca Polska*) have been impactful in gathering support of the wider community and accruing membership by accentuating the post-socialist experience of material and cultural dispossession (Kalb, 2005: 1). While this allows such ideas to become mainstream and/or co-opted by emerging right-wing elites, the ecologism of the far-right continues to be popularly denoted as eco-fascism, a socially outcast notion with unacceptable moral tenets. However, the incorporation of potentially problematic worldviews (naturalism, Manichaeism, and organicism), through claims to local ownership, into environmental issues, lobbying and bottom-up approaches, enables the proponents of FRE to be heard within local environmentalist networks, as the case of *Psia Ekipa,* an organisation with non-nationalist leadership (but some far-right membership) has shown. Thus, by reframing the debate around multifaceted notions of 'localism' and nativism, far-right ecologists have permeated local environmental networks.

This interaction irrevocably changes the ideological profiles and identities of those involved in the environmentalist assemblage. In order to cast away the eco-fascist label, the nationalist profile of some of these organisations (most notably, *Ecolektyw*) became assuaged and virtually invisible. Consequently, these ideological shifts and incongruities, paired with their decentralised organisation and lack of a clear operational agenda, led *Ecolektyw* and *Puszczyk* to become temporarily defunct, with its leaders either continuing their activism in far-right movements or focussing on their personal careers. Although attributing this exclusively to their interactions with external actors would be difficult, dropping the nationalist credentials along the transitional period has certainly contributed to their demise. Nevertheless, by engaging with (and co-constructing) these networks, the leaders of these organisations did not cease to be far-right nationalists. Moreover, being a far-right nationalist did not stop them from being environmentalists, nor did it undermine the motivation for integrating their worldviews into those of environmental activists.

Simultaneously, amid its initial defence, the environmental movements have, even if timidly, brought these actors on board through 'topical' collaborations, such as issues of animal welfare or deforestation. Ignoring the existence of this ethical dilemma on behalf of local activists renders the normalisation of far-right individuals and organisations conducive to a reassembling and strengthening of grassroots environmentalism. This attempt at normative broadening, which was achieved through the expansion of who counts as an 'acceptable' actor within mainstream environmentalism, points to the interactions occurring as a consequence of environmental issues and causes reflecting deep social fissures. That 'empowerment of the under-privileged' can also be co-owned by the far right signals the blurred lines of the normative component in politics and the environmentalist 'umbrella', but also of bottom-up projects against both the status quo and various forms of exclusion. Therefore, instead of spreading moral panic to fend off FRE by using the exact Manichaeism of the far right, perhaps it is time to embrace the ideological heterogeneity and contingency of environmentalism, and to re-orientate our attention towards the sources of discontent that introduce the 'extreme' within local networks for a more careful framing of grassroots activism and its goals.

References

Anshelm, J., and Hultman, M., 2014. A green fatwā? Climate change as a threat to the masculinity of industrial modernity. *NORMA: International Journal for Masculinity Studies*, 9(2): 84–96. https://doi.org/10.1080/18902138.2014.908627.

Barcena, I., Ibarra, P., and Zubiaga, M., 1997. The evolution of the relationship between ecologism and nationalism, in: Redclifft, M., and Woodgate, G. (Eds.), *The International Handbook of Environmental Sociology*. Edward Edgar Publishing, Cheltenham, pp. 300–15. https://doi.org/10.4337/9781843768593.00030.

Biehl, J., 2011. 'Ecology' and the Modernization of Fascism in the German Ultra-right, in: Biehl, J., and Staudenmaier, P. (Eds), *Ecofascism Revisited: Lessons from the German Experience*. New Compass Press, Porsgrunn, pp. 1–13.

Borras, J., 2019. Agrarian social movements: The absurdly difficult but not impossible agenda of defeating right-wing populism and exploring a socialist future. *Journal of Agrarian Change*, 20(1): 3–36. https://doi.org/10.1080/09644016.2013.755005.

Blühdorn, I., 2013a., The governance of unsustainability: ecology and democracy after the post-democratic turn. *Environmental Politics*, 22 (1): 16–36. https://doi.org/10.1080/09644016.2013.755005.

——, 2013b. *Simulative Demokratie: Neue Politik nach der postdemokratischen Wende*. Suhrkamp, Frankfurt. https://doi.org/10.5771/0032-3470-2014-1-169.

Blühdorn, I., and Welsch, I., 2007. Eco-politics beyond the paradigm of sustainability: A conceptual framework and research agenda, *Environmental Politics*, 16 (2): 185–205. https://doi.org/10.4324/9781315868998-5.

Bramwell, A., 1985. *Blood and Soil: Richard Walther Darré and Hitler's "Green Party"*. Kensal Press, London.

Bruggemeier, F. J., Cioc, M., and Zeller, T. (Eds.), 2005. *How Green Were the Nazis? Nature, Environment, and Nation in the Third Reich.* Ohio University Press, Athens, OH.

Castelli Gattinara, P., and Froio, C., 2018. Getting 'right' into the news: grassroots far-right mobilization and media coverage in Italy and France. *Comparative European Politics*, 17(1): 738–58 https://doi.org/10.1057/s41295-018-0123-4.

Castoriadis, C., 1998 [1975]. *The Imaginary Institution of Society*. MIT Press, Cambridge, MA.

Coupland, P., 2017. *Farming, Fascism and Ecology: A life of Jorian Jenks*. Routledge, Oxford.

Dawson, J. I., 1996. *Eco-Nationalism: Anti-Nuclear Activism in Russia, Ukraine, and Lithuania*. Minnesota University Press: Minneapolis, MN.

Dahmer, A., 2019. Pagans, Nazis, Gaels, and the Algiz Rune: Addressing Questions of Historical Inaccuracy, Cultural Appropriation, and the Arguable Use of Hate Symbols at the Festivals of Edinburgh's Beltane Fire Society. *Temenos, Nordic Journal of Comparative Religion*, 55(1): 137–55. https://doi.org/10.33356/temenos.83429.

Forchtner, B., 2019a. Nation, nature, purity: extreme-right biodiversity, in: *Cultural Imaginaries of the Extreme Right. Special Issue of Patterns of Prejudice*, 53(2), 11–31. https://doi.org/10.1080/0031322x.2019.1592303.

——, (Ed.), 2019b. *The Far Right and the Environment: Politics, Discourse, Communication*. Routledge, Oxford.

Forchtner, B., and Kolvraa, C., 2015. The nature of nationalism: Populist radical right parties on countryside and climate, *Nature & Culture*, 10(2): 199–224. https://doi.org/10.1080/0031322x.2019.1592303.

Freeden, M., 2013. The Morphological Analysis of Ideology, in: Sargent, L., Freeden, M., and Stears, M. (Eds.), *The Oxford Handbook of Political Ideologies*. Oxford University Press, Oxford, pp. 148–74.

——, 1996. *Ideologies and Political Theories*. Oxford University Press, Oxford.

Gare, A., 1995. *Postmodernism and the Environmental Crisis*. Routledge, London.

Griffin, R., 1991. *The Nature of Fascism*. Pinters Publisher Limited, London.

Harper, J., 2018. Introduction: Illiberal, aliberal, anti-liberal?, in: Harper, J., (Ed.), *Poland's Memory Wars: Essays on Illiberalism*. Central European University Press, Budapest and New York, pp. 1–22. https://doi.org/10.5771/9783828871847-1.

Humphrey, M., 2013. Green Ideology, in: Sargent, L., Freeden, M., and Stears, M. (Eds.), *The Oxford Handbook of Political Ideologies*. Oxford University Press, Oxford, pp. 496–514. https://doi.org/10.1093/oxfordhb/9780199585977.013.0011.

Inglehart, R., 1971. The Silent Revolution in Europe: Intergenerational Change in Post-Industrial Societies. *American Political Science Review*, 65(4): 991–1017. https://doi.org/10.2307/1953494.

Kalb, D., 2005. Introduction, in: Kalb, D., and Halmai, G. (Eds.), *Headlines of Nation, Subtexts of Class: Working Class Populism and the Return of the Repressed in Neoliberal Europe*. Berghahn Books, New York, pp. 1–17.

Krange, O., Kaltenborn, B., and Hultman, M., 2018. Cool dudes in Norway: climate change denial among conservative Norwegian men. *Environmental Sociology*, 5(1): 1–11. https://doi.org/10.1080/23251042.2018.1488516.

Kuchler, M., and Bridge, G., 2018. Down the black hole: Sustaining national socio-technical imaginaries of coal in Poland. *Energy Research and Social Science*, 41(1): 136–47. https://doi.org/10.1016/j.erss.2018.04.014.

Lubarda, B., 2020. Beyond eco-facism? Far-Right Ecologism (FRE) as a framework for future inquiries. *Environmental Values*, 29(6): 713–32. https://doi.org/10.3197/096327120x15752810323922.

——, 2017. Polluting Outsiders: Green Nationalism as a Concept — Case Study: Latvia, in: Aydın, M., Pınarcıoğlu, N. S., Uğurlı Ö. (Eds). *Current Debates in Public Finance, Public Administration, and Environmental Studies*. IJOPEC, London, pp. 437–57.

Lockwood, M., 2018. Right-wing populism and the climate change agenda: exploring the linkages, *Journal of Environmental Politics*, 27(4): 712–32. https://doi.org/10.1080/09644016.2018.1458411.

Mikecz, D., 2015. Changing movements, evolving parties: The party-oriented structure of the Hungarian radical right and alternative movement. *Intersections: East European Journal of Society and Politics*, 1(3), pp. 101–19. https://doi.org/10.17356/ieejsp.v1i3.112.

Olsen, J., 1999. *Nature and Nationalism: Right-Wing Ecology and the Politics of Identity in Contemporary Germany*. Palgrave Macmillan, New York.

Olwig, K., 2003. 'Natives and aliens in the national landscape', *Landscape Research*, 28 (1), 61–74. https://doi.org/10.1080/01426390306525.

Smith, A., 1991. *National Identity*. University of Nevada Press, Reno, NV.

Scruton, R., 2012. *Green Philosophy: How to Think Seriously About the Planet*. Atlantic Books, London.

Sörlin, S., (1999). The articulation of territory: landscape and the constitution of regional and national identity, *Norsk Geografisk Tidsskrift — Norwegian Journal of Geography*, 53:2–3, 103–12. https://doi.org/10.1080/00291959950136821.

Staudenmaier, P., 2011. Fascist Ecology: The 'Green Wing' of the Nazi Party and its Historical Antecedents, in: Biehl, J., and Staudenmaier, P. (Eds.). *Ecofascism Revisited: Lessons from the German Experience*. New Compass Press, Porsgrunn, pp. 13–42.

Uekötter, F., 2006. *The Green and The Brown: A History of Conservation in Nazi Germany*. Cambridge University Press, Cambridge.

Voss, K., 2014. Nature and nation in harmony: the ecological component of far right ideology, European University Institute, Florence (Unpublished doctoral thesis).

Żuk, P., and Szulecki, K., 2020. Unpacking the right-populist threat to climate action: Poland's pro-governmental media on energy transition and climate change. *Energy Research and Social Science*, 66(1): 1–12.

Greenline Front Official Blog, 2018. *The Law of Blood*. http://greenlinefront.blogspot.com/2016/10/the-law-of-blood.html.

6. Contorted Naturalisms
The Concept of Romanian Nationalist Mountains[1]

Alexandra Coțofană

This chapter investigates how logics and claims of indigeneity work within Romanian social media to construct and authorise divides between insiders and outsiders. The analysis focuses on social and political discourses that permeate local understandings and experiences of neoliberalism, class, and race. The chapter analyses data from interactions and posts on Romanian social media groups. Here, the underlying trope reinforces a 'pure' Romanian identity that claims to be indigenous to the land and Christian Orthodoxy, with constantly having to defend the land and identity from outsiders—characteristics that are seemingly mirrored by the actions of a landscape understood as sentient and deliberative. This chapter delves into one aspect of current Romanian anti-Semitic imaginaries, by analysing how Romanian social media and the right-wing blogosphere blend esotericism with a xenophobic brand of nationalism. These e-spaces, particularly Facebook groups and pages, are important, as they have tens of thousands of followers, and their discourses, logics, and ideas have migrated more than once to mainstream media outlets.

The chapter unfolds as follows: first, some context is given for understanding the content of the blogs and social media pages chosen for analysis. This information introduces Romanian cosmologies and

1 This chapter is dedicated to Vintilă Mihăilescu, who left us with too many ethnographic questions to figure out all by ourselves, when he passed on 22 March 2020.

 https://doi.org/10.11647/OBP.0244.06

the political ecologies that these online spaces produce from disjointed fragments of national history, contemporary conspiracy theories, and revived far-right ideologies of the past century. Then, the chapter moves on to describe historical moments in Romania's state formation that have shaped the particular form of indigeneity analysed here. Further, the chapter focuses on the events interpreted by the right-wing blogosphere as evidence for a sentient landscape intent on fighting Western occupation, in an attempt to create a comprehensive cosmological understanding of contemporary events. Throughout the chapter, I use the term *occult* to signify hidden ways, intentional or not, to escape or side-step modern rationality and its logics, ranging from rituals that scholars often uncritically call *esoteric*, to logics invoked in contemporary conspiracy theories. Furthermore, until recently, the academic study of esotericism and the occult has mainly been an historical affair, meaning they have been treated as historical and discursive topics, rather than lived experience (Crockford and Asprem, 2018).

Analysing the blogs and social media pages *in toto* would be too vast a topic to fit within a book chapter. Instead, I focus on two moments interpreted very differently. One occurred in 2011, when an IDF helicopter crashed in the tall Romanian Carpathian Mountains and resulted in several fatalities of both Romanian and Israeli military personnel. While official government sources concluded that the crash was due to human error, the right-wing blogs and social media pages developed their own version of the incident, claiming that it was the mountain itself that was responsible for the crash. The blogs and social media pages maintained that the mountain materialised fog as old bearded men, who created a strong wind that crashed the IDF helicopter into the mountainside. The motivation of the mountains to do this was interpreted as a protest against Israelis, who are understood as a symbol of Western occupation. Even though it may seem that it is this very sentient landscape that works with the occult, if we turn our attention to how the Western 'dangerous other' is imagined, we quickly see that the occult is also imagined there, in the 'West'. The second moment stems from the winter of 2017, when over one hundred spiritual pilgrims were recovered by mountain rangers in an area not far from the 2011 IDF helicopter crash. They had arrived to observe the energetic pyramid of the rock formation called the Sphinx, which is said to only be visible every year on the 28 November at 4:45 p.m. The tall snow and blizzard stopped the pilgrims on their way, yet these events were not interpreted on the same social media pages as the mountain attacking the spiritual pilgrims, despite the similar manifestation of the weather.

Philippe Pignarre and Isabele Stengers (2011) think of practices of anti-capitalist protests as a form of militancy meant to achieve a certain type of imagined purity. Pignarre and Stengers' approach is a useful lens for this study, as their own theoretical reference to Marx serves as a double mirror for how a particular logic of the occult is constructed in the Romanian context. For Marx, in his analysis of capitalism, the bourgeoisie is "the sorcerer, who is no longer able to control the powers of the nether world whom he has called up by his spells" (Marx and Engels, 2017: 58). Simply put, Marx imagines capitalism in a Faustian sense, in that the bourgeoisie is sacrificed on the altar of their own pursuit of capitalism.

If the Carpathians are imagined as the walls surrounding a citadel, Romanians are imagined as a human wall defending Western civilisation from Eastern occupiers of all sorts, from empires to barbarian tribes. For the authors of schoolbooks in Romania, this becomes the explanation for Romania's inability to develop as much as its Western counterparts, but also the reason why the West has experienced such great accomplishments unhindered. This sacrificial destiny is meant to imbue students with a sense of pride, but also to ideologically externalise the causes of political failure.

Constructions and understandings of the occult available in the Romanian esoteric blogosphere suggest that the Carpathian Mountains have the power to morph into indigenous elders and to strike down an IDF helicopter. These interpretations could reveal much about the logics and claims of indigeneity currently at work in the mainstream political realm. The fact that this connection may not seem immediately apparent hints at the sorts of logics with which we usually operate. Modern rationality has a tendency not to engage with things it does not understand, including the occult in all its many forms, and thus fails to imagine the alternative worlds that their existence might create. The elements of which these worlds are made are what Stengers calls the "unknowns of modernity" (2011). It is one of these 'unknowns' that this chapter pursues in its analysis.

'Nationalist' Mountains

Doing research for this chapter allowed me to see *home* with completely different eyes. While some say that anthropology starts at home, in my case, anthropology made a full circle and came back home, as I was doing research on Romanian politics and its entanglements with esotericism and discourses of the occult, and discovered with new eyes the Carpathian Mountains, specifically the Bucegi Massif, very close to where I was born and raised. As often happens, I believed home was benign, too banal to provide material for research, and that research was to be found elsewhere.

Fig. 1. The Fangs (or Sharp Rocks) of the Buck, where the Israeli helicopter crashed. Photograph by Alexandra Cotofana (2010).

The reality proved to be different: by monitoring social media platforms for discourse analysis on esotericism and the occult, I found articles about both the 2011 helicopter crash and the 2017 rescue mission of the spiritual pilgrims. These articles had very different interpretations of meteorological events that caught humans in their wake. The basis for these interpretations seemed to be a form of digital ethnic nationalism, where the presence of the IDF in the mountains was considered colonial, occult, malevolent, and thus their sacrifice was cheered on through invocations of anti-Semitism. On the other hand, when over a hundred people, mostly Romanian or sharing a certain understanding of the Bucegi Mountains as sacred, endangered their own lives by ignoring meteorological cautions, the same social media groups interpreted the events as just a meteorological misfortune, with no understanding or invocation of a vengeful, sentient landscape in sight. In short, the imaginary of a militarised, sentient landscape is only invoked when the trespasser is an ethnic Other who can be historically contextualised—in this case, by Romanian anti-Semitism.

Once my attention was redirected to a familiar place in an unfamiliar way, I started noticing more and more that people from home, including close family, distant relatives, neighbours I knew, and fellow Romanians I did not, were sharing links to the same articles related to these events on their Facebook pages. The articles were hosted by a few websites dedicated to esotericism and the occult and they embraced the latter theory, where the Bucegi Massif manifested elders that crashed the IDF helicopter, as punishment to the foreigners in it who were "trying to occupy ancestral land".[2] Comments from several relatives and neighbours revealed that many of them believed the old men who manifested from steam or clouds were Dacians, a pre-Christian population generally accepted by local historians as indigenous to the lands that make up modern-day Romania (Boia, 2001).

Of the social media and blogs available, I focus here on two in particular, because of their following and the popularity of their articles in Facebook shares. One website, the name of which translates to "know the world" (www.cunoastelumea.ro), has a Facebook following of 66,391; the second, called www.efemeride.ro, has a Facebook following of 84,999[3] people, and publishes highly trafficked articles starring Zalmoxis, a Dacian god, claiming that the Dacian culture is the oldest in the world, that they were the first inventors of an alphabet, that they had the only time- measuring tools for a certain historical period, that the great biblical flood took place in the west of the Black Sea, and finally, that Mount Olympus is actually Bucegi Massif, the mountain range

2 www.nationalisti.ro/2016/08/paranormal-un-elicopter-israelian-s-lichefiat-timpul-unui-experiment-secret-muntii-bucegi/.

3 Numbers valid for April 3, 2020.

in my hometown.[4] The Bucegi Massif is also famously believed to be Mount Kogainon, the holy mountain of the Dacians (Damian, 2019), which would explain the wrath of the elders who materialised from the mountains to attack the IDF helicopter.

The belief that Zalmoxis and Dacians are the oldest, most sacred of people, is a neopagan spiritual movement called Zalmoxianism. The movement blends pseudo-historical assumptions about the Dacians and their sacred spaces, such as the Omu Peak in the Bucegi Mountains, considered to be the main sanctuary and the most important energetic centre of the planet; with imported right-wing conspiracies, which act as an explanation for why the secrets of their greatness are not revealed to the world. On these social media pages, the culprits are the New World Order, and the occult Jewish free-masons, led by figures like George Soros and the Rothschild family. Zalmoxianism is a form of heathen Reconstructionism of a presumed old tradition, yet the written sources that can be used to historicise Dacian rituals are few and far between. The ancient historians Strabo and Herodotus are important for Zalmoxianism, as they both mention that Zalmoxis used a cave to retreat after he met with the Thracian elites in his role as high priest (Damian, 2019). Meanwhile, alternative histories have focused profusely on the idea that there is a series of underground caves under Bucegi that are used by communists, global occult elites, or whoever is of interest at the moment.

It became important, through this new research, to learn to re-consider people I had known for decades. What had happened? Had there been a change that my time spent abroad for my doctorate prevented me from seeing? Had people from my hometown always had these beliefs, and was I just now noticing? These questions were even more pressing as most re-posts from occult and esoteric blogs were from people who were also long-time voters for PSD, the centre-left-wing party that has dominated Romanian politics since 1989. The dominant PSD is the largest party in Romanian Parliament, and has the largest number of mayors, local and regional administrators. From the point of view of a political anthropologist, it seemed strange and baffling, at first sight.

---

4 https://www.efemeride.ro/un-mare-secret-mondial-ascuns-despre-romania-se-schimba-istoria-tarii-noastre.

Yet after a quick review of political events in Romania through the last three decades, matters became a bit clearer. After 1989, interest in Christian Orthodoxy, the imagining of a pure Romanian nation and the Dacians, who were a pre-Christian population generally accepted by local historians as indigenous to the lands that make up modern-day Romania, was the subject of discourse for a few far-right politicians and sometimes appeared in the discourse of centre-right parties PNȚCD (*Partidul Național Țărănesc–Creștin Democrat*, The Christian Democratic National Peasants' Party) and PNL (*Partidul Național Liberal*, The National Liberal Party; Alexe 2015). This all changed in 2014, when the traditional left-wing party PSD's (*Partidul Social Democrat*, The Social Democratic Party) presidential candidate Victor Ponta employed Romanian-ness and Christian Orthodoxy as central to his campaign against PNL candidate Klaus Iohannis, a Lutheran Saxon from Transylvania. Even though Ponta did not win in 2014 (or perhaps because of that fact), PSD's voters echoed the candidate's ideological focus in trying to make sense of national and international political affairs.

These blogs represent the area where I grew up—the Carpathian Mountains—as embodying the will of an ethnocentric nation. Below, I explore the elements that underlie the logic of these blogs and what makes their discourses believable to the tens of thousands of Romanians who follow them. Romania's nationalism, in its current form, is constructed from elements of xenophobia, fear of ethnic and religious 'Others', programs of governing that have supported, throughout the twentieth and nineteenth centuries, understandings of a Christian Orthodox indigeneity versus its many dangerous 'Others'. In 1918, a series of international treaties allowed for the formation of Greater Romania. As a consequence, the country doubled in size, with much of its new population made up of ethnic minorities. The new country might have been sustainable if the existing logic of ethnic nationalism had shifted to civic nationalism. Instead, the state encouraged the Romanian peasantry, the majority of the population, to help the new nation-state's development, by becoming urbanised, educated, and forming the new middle class. In doing this, and increasingly overlapping Romanian identity with Christian Orthodoxy, the state displaced an already urbanised middle class, the majority of whom were Jewish (Oncioiu, 2016). Yet this was not the first or last time Romania placed religious others in opposition to its Christian Orthodox national identity.

The Nation as Sacred

Chronologically, the Romanian Orthodox Church has existed for longer, and in a far more consistent institutional form, than the Romanian state. In fact, out of all the elements that form the Romanian national and ethnic collective imaginary, the discourse around Christian Orthodoxy presents the faith as being the oldest identitary element for the nation. The official position of the Romanian Orthodox Church recounts that after the apostolic synod from 49–50 AD, Andrew the Apostle had several missionary journeys. He started his fourth and final journey by preaching the gospel, travelling northward to Kyiv, Ukraine and then to Novgorod, Russia (Bandak and Jørgensen, 2012).

Fig. 2. St Andrew's Cave in Dobrogea, Southern Romania. Photograph by Statache Marian (2014), Wikimedia, CC BY-SA 3.0 RO, https://commons.wikimedia.org/wiki/File:Pestera_Sf._Apostol_Andrei_1.jpg.

He later reached Scythia Minor (now Dobrogea, a province in south-eastern Romania), where he remained for twenty years. The clerics I interviewed in the summer of 2017 claimed that the Apostle chose to remain on what later became Romanian territory for so long because of an affinity he developed towards the land and its people—a form of blessing in itself. Using this event to justify the discursive legitimisation of Romanianness in juxtaposition to all surrounding populations, Christian or not, serves to sacralise not only the space, but also the people, setting them apart from Slavic and Hungarian neighbours who were Christianised substantially later. This sacralised element of collective identity is centuries older than the first statist mention surrounding the formation of *Romanieness*. At the same time, the national myth is constructed around a number of purified pre-Romanian identity myths meant to offer cohesion and historical continuity to the historical self of the country.

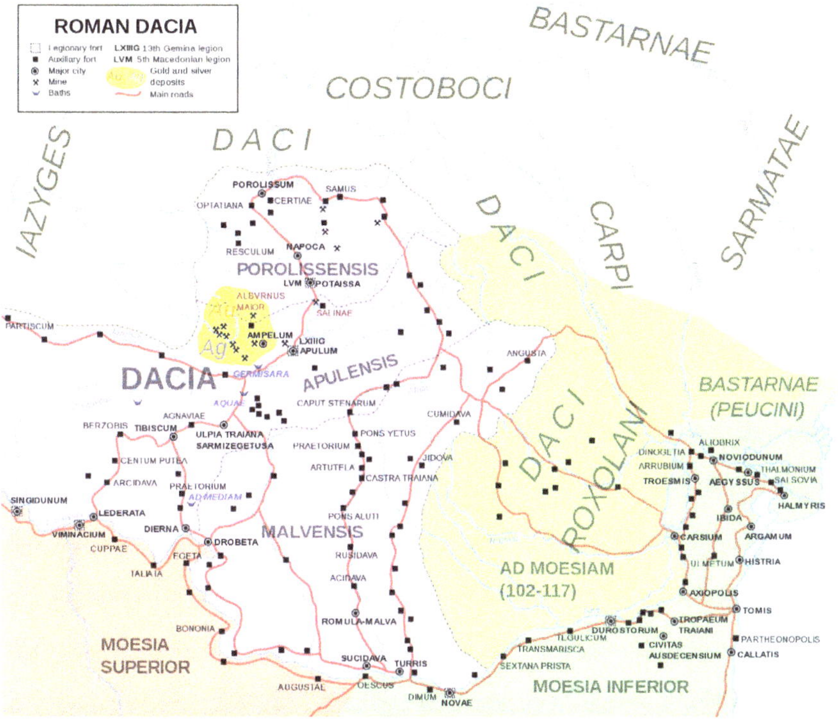

Fig. 3. The Roman Empire in the first century AD. Photograph by Hpdeparture (2019), Wikimedia, CC BY-SA 4.0, https://upload.wikimedia.org/wikipedia/commons/4/45/Dacia.png.

Historic accounts argue that the local Dacian population was conquered by the Roman Empire, and the newly formed province of Dacia Felix slowly morphed, linguistically and culturally, into today's Romania (Leuștean, 2007). The historical period of Dacia Felix is 106 to 274–75 AD (Boia, 2001), which means Romanians could claim they were Christian some fifty years before their national identity began to form, even by the earliest historical accounts. Together with a shared language, much of the Romanian national identity myth overlaps with Christian Orthodoxy (Stan and Turcescu, 2007). This discourse has historically been used as a grounds for claiming territories, forming the nation-state, and developing narratives of the collective self, showing endurance in the face of many religious colonial others.

Something that deserves particular attention here, for the purpose of our analysis, is the fact that several colonial Others have been identified by recent Romanian nationalist projects, before and after 1989, as either essential to the making of the Romanian collective identity, or, on the contrary, as intentionally delaying and sabotaging any projects of national unity. Romanian pupils before 1989, as well as in the decades after the revolution, learned through the country's public schools that the Roman Empire's conquest of Dacia, while essentially colonial, was fundamental to the making of the Romanian nation.

The same public-school system that has had a monopoly over Romania's K-12 curricula even after 1989, has maintained for decades that the Ottoman Empire's colonial project was the most destabilising to Romania's nation-making. As expected, this bias has been heavily used in recent anti-Muslim populist propaganda,[5] from legitimising the presence of Romanian soldiers in Afghanistan and Iraq, to justifying a political refusal to accept Syrian refugees as part of an EU member-set quota in recent years. A second rapacious, colonial Other in these discourses is the Austro-Hungarian Empire and their historical sovereignty over Transylvania,[6] followed by the vilification of various Slavic state formations, traditionally in the north-east of Romania's current borders.

---

5 https://www.dw.com/en/romanian-tabloids-incite-panic-over-refugees/a-40753654.)

6 https://hungarytoday.hu/angry-romanians-break-into-uzvolgye-ww1-memorial-site-attack-peaceful-hungarian-protesters/.

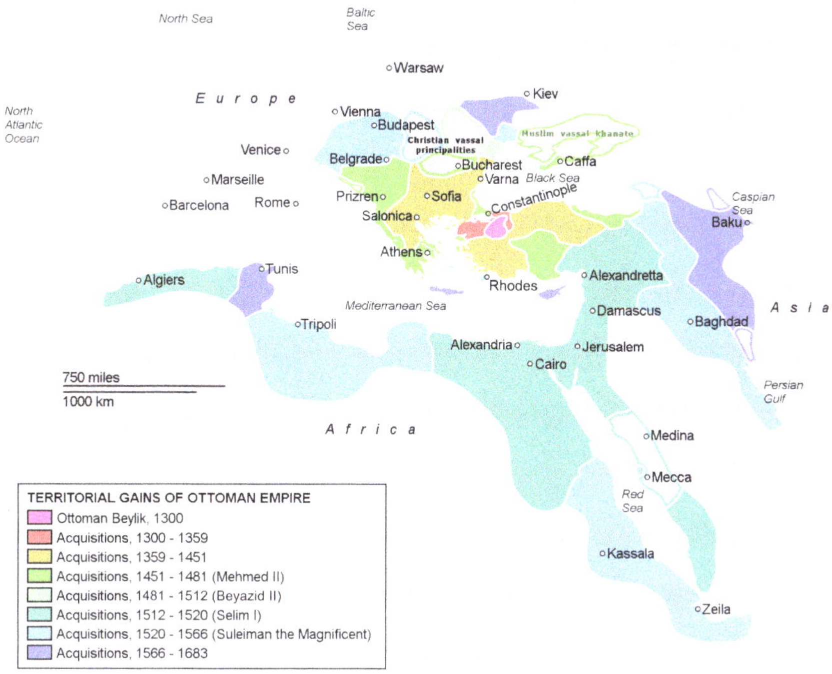

Fig. 4. The Ottoman Empire in 1683. Photograph by Atilim Gunes Baydin (2011), Wikimedia, Public Domain, https://commons.wikimedia.org/wiki/File:OttomanEmpireIn1683.png.

In the case of the Slavic state formations, the acidity of the discourse has been diluted by virtue of a shared Christian Orthodox identity, even while Romanian Orthodoxy is perceived as closer to Greek Orthodoxy than the Orthodoxy of the Slavic populations—at least when the colonial project was exported as atheistic, as in the case of the Soviet mode (Tismăneanu, 2003). Most of these hostilities are born as a result of (arguably imposed) international treaties that have allocated territories against one national interest or another, signed throughout the last two centuries. The undesirable colonial Others are depicted in Romanian nationalist discourse as having made illegitimate claims to Romanian territories since time immemorial.

Even though the history of the Ottoman Empire is extremely long and politically complex, Romanian pupils learn very early in their history classes that the Romanian principalities were the "gate of Christianity" (understood as a symbol of keeping intruders out) and that Romanian leaders were the "keepers of Christianity" (Boia, 2001: 67), with the Carpathians and the Bucegi Massif playing a central role in this discourse.[7] This nationalist discourse relies on the narrative that Romanians, even in their earlier divided administrative forms, had to face the Ottoman threat alone—and somehow succeeded—without much aid from the Western Christian countries that they also, in effect, protected.

Lucian Boia's clever analysis of the ideological pitfalls of the Romanian reconstruction of the past deserves to be quoted. His 2001 *History and Myth in Romanian Consciousness* and his 2004 *Romania* (*Topographies*) help deconstruct and explain some of the rather improbable stories that public Romanian education reproduces *ad infinitum*, without any seeming efforts to reform the system in order for pupils to learn a more nuanced history, which would better reflect historical complexities. Boia explains how the hyper-focus on a small number of victories against the Ottomans (and further others, legitimately or illegitimately perceived as colonial) has led to a construction of the past where Romanians managed time and again to defend their land, and the whole of Christianity, almost supernaturally, when everyone around was losing. In itself, this making of a superhuman past could be discussed as occult.

The Ottoman Empire is imagined as the longest-standing outside threat, and an unwelcome form of colonisation. It is the main focus of most history books available in Romanian K-12 education, while at the same time, a myriad other "invasions" of "migratory tribes" (Almaș, 1987) are all briefly mentioned and lumped together during primary and secondary education. This last move is in no way accidental—by treating the existence of Gepids, Huns, Avars, pechenegs, Cumans, Oghuz, Alans, or Tatars, to name just a few, as migratory tribes with brief and unimportant interventions on Romanian territory between the fourth and twelfth centuries BC, the public education system

7 https://adevarul.ro/locale/buzau/tunelurile-secrete-romaniei-aparut-triunghiul-aur-daciei-reteaua-bucegi-egipt-tunelul-piramida-paranormala-ceahlau-1_5515963e448e03c0fdf87dad/index.html.

emphasises the main point of its national myth: *We* were already here, *They* were passing through.[8] The image of the Ottoman conqueror is different to that of the Roman Empire: partly as a result of their shared, pre-1054 schism, Christian identity, the Romans have been depicted as an 'accepted' colonial other. They were imagined as the Western, civilising half to the Dacian, local half of language and identity that make up the nation (McGuckin, 2010). This myth of the 'civilising West' continues today, albeit in different forms.

Efforts to manipulate the past remind many of the cultural project started in the early 1970s by the Ceaușescu regime. This project intended to "purify" national identity, as part of a broader political project of sovereignty and independence from the USSR and other world powers (Boia, 2001). While the goal in itself sounds noble, and, as academics fluent in the language of postcolonialism, we are taught to support the plight of small countries attempting to rid themselves of the chains of domination, the example of Ceaușescu's method has one major flaw. As I will discuss, one of the main tools for Ceaușescu's project of producing historical purity was through academia. While many scholars were forced to obey new sets of rules and cultural policies, others enthusiastically joined in the making of the national myth.

This went as far as a team of archaeologists discovering human remains in the area where the dictator was born and advancing these findings as *Homo oltenicus* (named after the region, Oltenia), claiming that they were the oldest human remains in Europe (Abagiu, 2007). As a symbol of the way Ceaușescu's cult of personality was built, the case of *Homo oltenicus* allows us to analyse complex subjectivities in the creation of the state and its mirroring in the human body. In no way historically central from an archaeological or political point of view, Oltenia is a region mostly remembered for having been the childhood home of Ceaușescu. Yet with the ideological creation of *Homo oltenicus*, Oltenia becomes, for its ideologists, equal to other important historical regions around the world, that lend their natural force to their most notable humans. In other words, this invented archaeological fact is meant to mimic some small-scale Egypt, Viking, or other racial myth that can later be successfully moulded into a populist project.

8 https://www.descopera.ro/cultura/2753237-dacii-niste-barbari-va-inselati.

Despite these contested forms of broad ethnogenesis, meant to construct the Romanian state in the collective memory as eternal, mono-ethnic and as legitimised by endurance, the state itself is fairly young, having only been formed in 1859 (Istodor, 2015). However, there is a different institutional agent that can act as a source of validation for this particular type of imaginary of indigeneity. The Romanian Orthodox Church has a longer history and has been both a source and object of significant argument in international political negotiations, involved in the redrawing of borders throughout history. It kept its privileged position, even under rule by the Romanian Communist Party (Kovacevik, 2008). The Church has maintained its importance as an element of national identity after 1989 (Racu, 2017), and even while some of its practices are publicly contested (Tismăneanu, 2003), it remains a core political, economic, and social actor in Romania. The Church is semiotically elastic in Romanian politics, as it has been used equally by the far-right Iron Guard during WWII, as a subtle but firm indicator of nationality in socialist history books, and by the centre-left PSD after 2014 (Racu, 2017).

Xenophobic Sentient Landscapes

From Pignarre and Stengers' point of view, the Carpathian incarnation of anti-capitalist protests that hit down an IDF helicopter as a symbol of Western domination, could be seen as militant, in service of a certain imagined purity (2011). In this case, the occult is used on both sides of the conflict: on the one hand, the Israeli military are portrayed as acting towards the fulfilment of plans set up by global elites, with a hidden political agenda, and at the same time, for attempting to enter Romanian national space secretly, by landing their military helicopter on the mountains, which is subsequently viewed and depicted as an act of capitalist sorcery, as Pignarre and Stengers would most likely agree.

Can landscapes be mobilised and re-imagined as defending national interests, and how would such a way of being be imagined? In Romania's case, the idea that the local Romanian ethnic population and the forest are 'brothers' has been capitalised on for the making of a unified ethnic nation since at least the creation of Greater Romania in 1918 (Mehedinți, 1927). Yet is the mountain imagined in this local ontology as *sentient*?

Does this *sōmaton* (bodied thing, thing with a material body) this being with a physical body, have life in it? In many ways, the mountain could be imagined as the exact opposite of the *daimōn* (spirited thing): while the mountain has a very dense body, we do not imagine him as possessing a soul. At the same time, while the *daimōn* is an *asōmaton* (thing without a physical body), we do consider it to be sentient, and to act according to its life source. In this theoretical analysis, it might be the very arrangement of landscapes as occult in the minds of the locals that aid in this project.

More than anything, the power of governing with the occult in Romania is kept alive by myths promulgated by politicians through social media outlets, which appeal to the masses and to some very specific anxieties. For example, some of the articles that received the most traffic on social media pages (and affiliated websites or blogs) imagine the Romanian national space in very specific ways. In trying to explain why the mountains may have attacked the helicopter, articles on these websites made note of the fact that the mountains were protecting a secret entrance in a complex system of tunnels within the mountain. Based on the source, these tunnels are either said to have served the Dacians or to have served the occult, hidden forces governing the world today. In this second sense, the mountain would lose any sentient quality within the local ontology.

Furthermore, the network of tunnels might be made by humans, be they forefathers or our contemporaries, or by aliens, according to some.[9] This, once again, shows that the idea of governing with the occult is a conglomerate made up of many disparate pieces, which cannot be connected easily.

If anything can be illuminated, it is that most of these theories became popular particularly after an author, Radu Cinamar, started publishing books about the 'occult ways of governing' in Romania and their locales. According to some, Radu Cinamar is the pseudonym of a group of former secret agents. According to other theories, it is the pseudonym of a female writer called Dorina Chirilă, who lives in north-western Romania. Either way, the interest in the Bucegi Mountains as a space for occult governing spiked after Radu Cinamar's 2003 book,

9 http://www.redescoperaistoria.ro/2015/03/29/tunelurile-secrete-ale-romaniei/.

Fig. 5. The Sphinx, one of the many rock formations atop the Bucegi Massif, linked to occult beliefs. Photograph by Sorin Besnea (2013), Wikimedia, CC BY-SA 3.0, https://upload.wikimedia.org/wikipedia/commons/thumb/0/0e/Sfinx_-_Bucegi.jpg/800px-Sfinx_-_Bucegi.jpg.

Future with a Skull (*Viitor cu cap de mort*), which claims there is a secret entrance in the Bucegi Mountains protected by energetic walls.

One of the reasons why parts of this ontology may seem foreign to some readers stems from the fact that, as Phillipe Descola reminds us, categories of interiority and physicality tend to derive from the "universal experience of being an intentional subject with a body" (2007). As such, we can only envision bodies that are like ours, to be able to produce things that we imagine as unique to our consistency, such as emotions. As a consequence, we find it hard to believe that *sōmata* with a denser consistency than the fleshy body of mammals, such as rocks and mountains, might be able to experience life in similar ways to us. In an intellectual exercise similar to Descola's observation around how technology moved reindeer herders from animism to analogism, could we think of the alleged tunnels in the Bucegi Mountains, assuming they exist, as a technological element that, regardless of who built them, help us to understand ways of being as a body of stone? Could we think through life (can we even imagine there being any?) as a stone *sōmata*? And if so, what would that look like? How do they manage to materialise fog into elders, and how do they, at the same time, protect a hyper-technological centre from the knowledge of the many?

A useful term here comes from Elizabeth Povinelli, who coined the concept of 'geontologies'. Povinelli compares geontology to Foucault's biopolitics, and claims that geontology "does not operate through the governance of life and the tactics of death but is rather a set of discourse, affects, and tactics used in late liberalism to maintain or shape the coming relationship of the distinction between Life and Nonlife" (Povinelli, 2011). Overall, 'geontologies' as a term might have potential for displacing Western metaphysics and allow some room for other ways of containing Life or Nonlife definitions and ways of being as they are discussed and conceived in academic circles.

How we order what counts as Life and what counts as Nonlife is a matter of cultural perspective, and a study on the ontology of rock formations as sentient would be most welcome. So far, the cartographic features of eastern Europe, especially of its mountains, have received most attention during World War I and World War II, revealing struggles for power, political affinities and challenges to authenticity, yet never has attention focused on rock-life, creating instead what Yuliya Komska calls a "discontiguous eastern Europe" (2018). Komska says, when referring to eastern Europe: "The area's natives, we point out, have consistently forged links to discontiguous lands and populations, whether willingly or by force" (2018: 4). In this sense, Komska is using the term interchangeably—for land and politics alike, for tropes about a collective self, as well as for understanding historical eras.

In a 2017 edited volume, several Romanian journalists and social scientists published a series of essays critiquing historian Lucian Boia's 2012 book, *Why is Romania Different?* (*De ce este România altfel*). The critical edited volume plays on the title of Boia's book, and is called *Why is Romania this way?* (*De ce este România astfel?* 2017). Here, the morphological difference between "other" (*altfel*) and "this way" (*astfel*) is minimal, so it can be aptly used for the sort of intellectual exercise attempted here. The larger question revolved around why a renowned intellectual such as Boia, who has built a career on extremely valuable work that deconstructs the politically-driven manufacture of the Romanian past in the service of nationalism, would write a book that, in a sense, goes to disprove all of what came before? Why would one imagine Romania as 'different'? Furthermore, what does this difference mean, what form does it take?

The title of Boia's book suggests an already assumed truth that Romania is different—rather than question whether or not this may be the case. The subject became a widespread topic of criticism as it signified that a preeminent analyst of populism had been seduced by a populist-related discourse of exceptionalism. Boia's critics reminded him that exceptionalism tends to create value judgments of social realities rather than describe them. Beyond this, to have this exceptionalism legitimated by the work of a historian such as Boia, who built a career on deconstructing nationalist discourse and practice in Romanian history, seemed a strange and perhaps dangerous direction.

This discourse of difference seems to serve a few purposes, all serving nationalism and certain brands of populism. On the one hand, it could easily be attached to discourses that claim Romania has been different *ab originem, illo tempore*, and that there is nothing the country can do about its perceived state of always lagging behind. And behind who? Behind the West, of course.

This question leads to a second goal—by claiming the assumed lagging in various economic, political and cultural aspects, this difference legitimises discourses that claim Romania is an internal 'primitive' of Europe, perpetually behind in terms of development (Mihăilescu, 2017). This, in turn, suggests a perpetual Romanian aspiration to become like the West, as well as giving rise to an apologist logic that suggests that the country is not headed in the 'right' direction (or quickly enough), due to the country's inherent difference. This apologism often eludes the speaker who, by virtue of their clear vision of the country's mentality, portrays themselves as a potential driving force, crippled by a national majority of internal European primitives.

The beginning of national identity-building for Romania began at a time when the two reference points were the 'West' and the 'East'. Rome, as an ideal Other, who rounds up the indigenous Dacian identitary half with a civilised Western half, is at the core of the fabrication of the Romanian identity, a purified genealogy developed as a strategy and justification against eastern powers (Vasile Pârvan, 2018 [1919])—this was later completed with the autochthonous Dacian identity, which legitimised claims for a local continuity (Mihăilescu, 2017).

This coupling of West and East historically had two consequences—on the one hand, the denial of Oriental connections, especially in the context of the neighbouring, expanding Ottoman Empire, and on the other hand, a hyper-emphasis on Western genealogies. Ernest Renan observes that civic nations, such as France, require their myths and use "strategic amnesia" in order to be able to survive as a nation (1992 [1882]). These stories of the self are what Sorin Antohi calls "ethnical ontologies" (2002). Ethnical ontologies have myths that can receive both good and bad interpretations. The most unexpected example in this sense is that of Romanians using the concept of the Self as European internal primitive, either to account for corruption, belligerence, alcoholism, or, on the contrary, to legitimate the myth of the noble Dacian, where belligerence and alcoholism are seen as natural traits of the noble, pastoral tribe.

According to Vintilă Mihăilescu, collective identity specific to small peasant nations in eastern Europe and the Balkans that emerged as post-colonies of empires drove several waves of political movement for independence in Romania: it was not individual freedom that was sought, but the collective independence of the group (2017). This collective self is what I call the *body of bodies*, the understanding of the state through the entirety of its people, versus thinking of the state as a cumulus of institutions. A good example of how this distinction might prove relevant is Alain Badiou's *Of an Obscure Disaster* (1991), where Badiou criticises Stalinism for having betrayed and perverted the ideal of communism by entrusting it to the state, instead of allowing what I call the *body of bodies* to remain the creative force.

This collective social body sought to differentiate itself from neighbouring nations by highlighting the uniqueness of its geography and of how 'The Romanian' (not one particular individual, but the mythical individual symbolically standing for the state) has always been at one with this nature, making sense of it in a way that only an autochthonous Self could.

In this sense, Romanians have a plethora of proverbs—the forest and the Romanian are brothers (*codru-i frate cu românu*), strong as a cliff (*tare ca stânca*), the forest whispered to someone (*i-a șoptit pădurea*), the man is like grass, his days are like flowers (*omul e ca iarba, zilele lui ca floarea*), the working man is like a tree giving fruit (*omul muncitor, ca pomu roditor*). Furthermore, the Carpathian Mountains are imagined as the core of the national space, even as they separate the country in two, and have historically been an efficient metaphor for imagining the country as mono-lithic, united through a common geological body. The mountain-as-national-being can be an important analytical tool for understanding nationalist ontologies.

This connection between man and natural space seeks to give the sense of a long-lasting collective identity—according to some historians, the people inhabiting these lands are "some of the oldest peoples in Europe" (Mehedinți, 1927). However, this geography has also been used to create a discourse of inescapability for the body of bodies. Caught between mountains, rivers and the sea, the collective body had to constantly sacrifice and be crippled by invaders. Furthermore, this assumed identity of autochthonism, intentionally contrasted with the labelling of all other ethnic or national identities that have interacted with the physical space as migratory invaders, meant that the collective body had to learn how to selectively legitimise political behaviours and discourses (Naumescu, 2011).

As we have seen, tropes about indigeneity are a culturally-productive category that can reimagine militarised relationships with more-than-human agents, mountains included (Carey, 2016). What sorts of rhetoric are being used to legitimise the landscape to act in anger? In the case of the Bucegi, local, massive, illegal (and portrayed-as-legal) deforestations are taking a toll on the health of the mountain. In the wood industry, neoliberal logics of economic growth, employment, and cheap furniture enable industry giants like IKEA to benefit, almost in the absence of any protest, from the wood of Europe's last old-growth forests, found in the Carpathians. This, together with the hidden, non-transparent business practices of local politicians, allows for the wood trade to remain dangerously unregulated.

To conclude, ethnographies of *sōmata*, as spirited, bodied non-humans have the potential to productively reconceptualise scholarship focusing on affective spaces, institutions (and implicitly, what we de-institutionalise), and theories of materiality and immateriality. Analysing their involvement with the occult could offer a critical outlook on the hierarchy of epistemic regimes at work in academia, as well as the possibility to re-think the porosity of conceptual borders. Comparing modes of thinking and logics that create both individual *sōmata*, such as demons, spirits, and institutional ones, such as the state, the church, economy, and so on, is an important place to start.

References

Abagiu, M., 2007. Protochronism and nationalism under Ceauşescu's brand in totalitarian Romania, *Arhivele Olteniei*, 21, 117–23.

Alexe, D., 2015. *Dacopatia şi alte rătăciri româneşti*. Humanitas, București.

Almaș D., 1987. Povestiri istorice pentru copii şi şcolari — şoimi ai patriei şi pionieri (I) Editura Didactică şi Pedagogică, București.

Antohi, S., 1999. *Imaginaire culturel et réalité politique dans la Roumanie moderne. Le Stigmate et l'utopie* ("Cultural Imagination and Political Reality in Modern Romania. The Stigma and the Utopia"). L'Harmattan, Paris-Montréal.

Badiou, A., 2010. *Of an Obscure Disaster.* Jan Van Eyck, Maastricht.

Bandak, A., and Jørgensen, J. A., 2012. Foregrounds and Backgrounds. Ventures in the Anthropology of Christianity, *Ethnos*, 77:4, 447–58. https://doi.org/10.1080/00141844.2011.619662.

Boia, L., 2001. *History and Myth in Romanian Consciousness*. Central European University Press, Budapest.

Carey, M., Jackson, M., Antonello, A., and Rushing, J., 2016. Glaciers, gender, and science: A feminist glaciology framework for global environmental change research. *Progress in Human Geography* 40(6), 1–24. https://doi.org/10.1177/0309132515623368.

Crockford, S., and Asprem, E., 2018. Ethnographies of the Esoteric. Introducing Anthropological Methods and Theories to the Study of Contemporary Esotericism. *Correspondences*, 6 (2), 1–23. https://doi.org/10.1163/9789004446458.

Damian, C. I., 2019. (Re)Inventing Sacred Places in the Context of Contemporary Paganism. Stonehenge and Kogaionon. *European Journal of Science and Theology*, 15 (3), 49–60. https://doi.org/10.33051/1841-0464-2019-15-3.

Descola, P., 2007. Apropos de Par-delà nature et culture *Traces. Revue de Sciences humaines*, 12, 231–52. https://doi.org/10.4000/traces.229.

Istodor, G., 2015. Transcendent and Immanent in the Orthodox Theology. *Dialogo*, 2, 45–54. https://doi.org/10.18638/dialogo.2015.2.2.4.

Kovačević, N., 2008. *Narrating Post/Communism. Colonial Discourse and Europe's Borderline Civilization*, Routledge, London. https://doi.org/10.4324/9780203895252.

Komska, Y., Moyd, M., and Gramling, D., 2018. *Linguistic Disobedience: Restoring Power to Civic Language*. Palgrave Pivot, London. https://doi.org/10.1007/978-3-319-92010-8.

Leuștean, L., 2007. Between Moscow and London: Romanian Orthodoxy and National Communism, 1960–65, *The Slavonic and East European Review*, 85(3), 491–521. https://doi.org/10.1057/9780230594944_8.

Marx, K., Engels, F., 2017. *The Communist Manifesto*. Pluto Press, Northampton. https://doi.org/10.2307/j.ctt1k85dmc.

McGurkin, J. A., 2010. *The Orthodox Church: An Introduction to its History, Doctrine, and Spiritual Culture*. Wiley-Blackwell, New York.

Mehedinți, S., 1927. *Școala română și capitalul biologic al poporului roman*. Biblioteca eugenică și biopolitică a ASTREI, Cluj.

Mihăilescu, V., 2017. De ce este România astfel' Avatarurile excepționalismului românesc. Polirom, București.

Naumescu, V., 2011. Postsocialist Europe: anthropological perspectives from home. *Journal of the Royal Anthropological Institute*, 17(3), 648–49. https://doi.org/10.1111/j.1467-9655.2011.01712_16.x.

Oncioiu, D., 2016. Ethnic Nationalism and Genocide: Constructing "the Other" in Romania and Serbia, in: Ümit, Ü. U. (Ed.), *Genocide*. Amsterdam University Press, Amsterdam, pp. 27–48. https://doi.org/10.1515/9789048518654-003.

Pârvan, V., 2018. *Datoria vieții noastre*. Școala Ardeleană, Cluj.

Pignarre, P., and Stengers, I., 2011. *Capitalist Sorcery. Breaking the Spell*. Palgrave Macmillan, London.

Povinelli, E. 2016. *Geontologies: A Requiem to Late Liberalism*. Duke University Press, Durham, NC.

Racu, A., 2017. *Apostolatul Antisocial. Teologie și neoliberalism în România postcomunistă*. Tact, București.

Renan, E., 1992. What is a Nation?, in: Renan, E., *Qu'est-ce qu'une nation?* Presses-Pocket, Paris. https://doi.org/10.1522/030155027.

Stan, L., and Turcescu, L., 2007. *Religion and Politics in Post-Communist Romania*. Oxford University Press, Oxford. https://doi.org/10.1093/acprof:oso/9780195308532.001.0001.

Tismăneanu, V., 2003. *Stalinism for all Seasons. A Political History of Romanian Communism*. University of California Press, Berkeley, CA.

7. A (Hi)Story of Dwelling in a (Post)Mining Town in Romania

Imola Püsök

For many people in Roșia Montană, living within *their* environment means mining gold, whilst also attending to some farm animals, a small garden, and collecting hay in a fairly traditional manner on the steep slopes surrounding the villages. The process of dwelling imbues the landscape with aesthetic qualities that prompt the unfolding of what Michael Uzendoski calls "somatic poetry": webs of interwoven lives and life narratives emergent in complex and dynamic social and ecological relationships, the experiencing of beauty and belonging (Uzendoski, 2008). In the somatic poetries of people in Roșia Montană gold and mining have central importance—primarily as indicators of worth and value—and experiencing and articulating beauty and value differently isolates locals from 'Others' and emphasises generational differences in understandings of loss.

In this chapter I present a (hi)story of dwelling in Corna, a village that belongs within the commune[1] of Roșia Montană. My aim is twofold: to unveil the context, i.e. the micro-world where individual stories and the lives of locals unfold and intertwine, on the one hand, and to weave my own analysis into the landscape on the other. In describing the context I draw on the narrative of a local councillor, *'nea* M.,[2] who in his very

1 A commune (*comună*) in Romania is an administrative unit consisting of several villages, which have a common mayor and a local council. The commune of Roșia Montană consists of sixteen villages.

2 *'Nea* is a short form of *nenea* (uncle). Although I would always address my interlocutors as *Domnul* (Mr) or *Doamna* (Mrs), in my writings I tend to refer to older neighbours with the more informal *'nea* and *tanti* (aunt), due to being introduced to the field through a young couple who had moved there in 2015. Although my

 https://doi.org/10.11647/OBP.0244.07

eloquent, persuasive and oratorial style, took me back to times to which all my other interlocutors made references, but never described in such a detailed and composite manner. I chose his narrative because he was the only one to present details, context and general information about Corna in the form of a '(hi)story' (*poveste*), and because many of these details and information—albeit in a different manner—would sooner or later resurface in the accounts and life stories of my other conversation partners. I will complement his story with my story-analysis. This latter story-analysis I understand to be the ongoing conversation between many of the voices (some of them more prominent than others) that I encountered throughout my research, both in person and in reading. I first started this research back in 2011, and then recommenced it in 2017. In the process of presenting this particular (hi)story of dwelling, I would like to draw attention to three aspects that emerged from my conversation with '*nea* M.'s story: temporality and livelihoods, social and ecological (inter)actions, and an emerging aesthetics within the telling of a (hi)story dwelling.

First, there are three distinct ways of experiencing time in the (hi) story of Corna, that surface from *nea* M.'s narrative, noticeable both in the ethnopoetic style[3] as well as in its content. These different ways of conceiving time and ways of living are defined by three successive political-economic regimes, delineated by macro-economic changes that severely impacted the way people could earn their livelihoods.[4] The

relationship with them will probably never become so close as to call them aunt and uncle in personal conversations, the terms used in my writings express the sense of familiarity I feel towards them in the field.

3 There are a total of eight verb tenses in the Romanian language, four of which are used for the past tense. The *imperfect* (past) tense marks cyclical actions, habits or actions that were not finished, while the other three mark actions that were finished at various points before the present or past. The narrative style I refer to throughout the text includes the rhythm of talking and the choice of actions that are related, but is also very much defined by the choice of verb tenses. For the pre-communist period '*nea* M. consistently uses the *imperfect* tense (marking cycles, or habits), when talking about the communist period, he always uses *perfectul compus*, which marks definite finished actions, and for the present—where the empirical uncertainties are more clearly defined through the lack of focus—he uses the present tense.

4 Although the three time constructs roughly correspond to Bauman's categorisation of societies, i.e. cyclical time for stable societies versus linear time for developmental societies; future-oriented time in modernism versus permanent present in postmodernity, or in liquid modernity (Tarkowska, 2006), I am reluctant to refer to this grouping, as I do not consider the past to be a different society (see Ingold et.al.,

pre-socialist period is marked by a cyclical time in how '*nea* M. recounts this period, which is manifest in the portrayal of repetitive, patterned activities, a perpetual rhythm of constant interaction with the landscape. The shifts in the way of living later presented in the story invoke changes in people's relationship to this land, to this landscape and to the state. While in the pre-socialist period miners relied on the landscape for a living, in the socialist era this turned into complete reliance on the state as the primary 'provider' of wealth and social life. Whereas the landscape is central to this way of dwelling in the sense that its 'voice' is constantly 'heard' in the conversation between people and their surroundings, later it becomes viewed as a setting and a tool for economic advancement, both for individual actors and for the state. The socialist era is then recounted in a linear style, where time becomes flattened, actions are completed and follow each other in a consecutive manner. Goals and life trajectories become visible, and focus from daily or annual cycles turn towards a vision of an attainable future. The third period depicted is the post-socialist period, wherein time becomes fragmented, life turns unstable and unpredictable, and the narrative style becomes erratic. It is a characteristic "liquid modernity" in its state of permanent *precarité* (Bauman, 2001: 41–42), and a "permanent liminality" in the way the ever-expanding, accelerating present is experienced as anti-structure (Szakolczai, 2015a; 2015b; Thomassen, 2015; Wydra et al., 2015; see also Eriksen, 2016). The process of dwelling—along with the landscape—is lost in civil and corporate discourses about the environment, about unemployment and poverty, about profit, wealth and exploitation, about sustainability and tourism (see also Velicu, 2015). The conversation that defined dwelling in the first part of the story has been reduced to a minimum: the landscape has become alienated by the mining company and environmentalist discourses, it has lost its giving properties in the eyes of many locals (who, unlike newcomers, do not see berries, healing plants or tourism as a valuable resource) and it has become increasingly threatening through its aggressive regenerative force, which confers an unfamiliar image to a mining landscape.

1996). However, I do refer to the (post-socialist) present as liquid modernity (see Bauman, 2000; 2001), which in my view has many similarities with the concept of permanent liminality in the way instability (flexibility, if you will) is made standard (see Horváth, 2013; Szakolczai, 2015a; 2015b; Thomassen, 2015; Wydra et al., 2015).

Second, I would like to point to the changing social and ecological dynamics in Corna. I have already noted the ways in which livelihoods for miners' families have changed in the three different economic structures, but I would also suggest that the changing landscape is a visual testament to a changing social fabric in the past twenty-three years of accelerating post-socialist transformations. The landscape, according to Ingold, is the "taskscape in its embodied form: a pattern of activities 'collapsed' into an array of features" (Ingold, 2000: 198). Social relationships, therefore, are intrinsic to the landscape in which humans dwell, as in the process of dwelling, "humans attend to one another" (ibid.: 196).

Finally, through my conversation with *'nea* M.'s narrative, I would like to allude to the aesthetics (of loss, of beauty, of belonging) emergent in telling the (hi)story of dwelling in Corna. In this I stress the inseparability of life and story (Uzendoski, 2008): the creative improvisation (Ingold and Halland, 2007) through which the historical time, the temporal perspective and the rhythm of storytelling are in unison, through which the different modes of livelihood are a metonymy for state-political processes, and through which the dynamics of the processes of dwelling become articulated. By following Uzendoski's (2008) idea of somatic poetry, I stress that the (hi)story of dwelling is not simply textual, it is an intersection of many lines (Ingold cited in Uzendoski, 2008: 13), a composite poetry that unfolds—almost implodes—in the many interactions it holds. In this respect my focus in determining the somatic poetries emergent in Roșia Montană differs slightly from that of Uzendoski, who in his account of two healing experiences in Amazonian Ecuador and Amazonian Peru, emphasises the "visceral role the body plays in experiencing [...] poetry" (ibid.: 25). The differences in our focus are also determined by the differences between the stories and experiences growing in the Amazonian rainforest and those unfolding in Roșia Montană: the differences between the texts and their respective contributive creative subjectivities, and the differences between our relationships with our respective research sites and participants. The lived and narrated somatic poetries of the people from Corna incorporate the Romanian macro-historical processes to the same degree as they include personal feelings and experiences. Instead of the bodily dimensions of experiencing and creating beauty, therefore, I will emphasise how the dynamics of interactions between the various participating agencies, both human and non-human, define the emergent (hi)story of dwelling in Roșia Montană. These interactions include the conversation with *'nea* M., the two women (his wife and our neighbour) watching us from the side, my many encounters with local people from Roșia and Corna, each of which informed and changed the way I could add to the story, and they also include the dynamic and changing interactions between people in the process of dwelling. These I could not personally witness, but rather, were already presented to me as a (hi)story. My conversation with *'nea* M.'s text, therefore, is also a conversation between the empirical realities and the various types of relations and networks I am part of during my field trips, and the many types of (hi)stories that are disclosed to me in conversations.

In the following section I will first describe the event-context which led to Roșia Montană becoming one of the places most frequently visited by social scientists and archaeologists. This is a context that strongly determines every story that is told in and about the area—including my own attempt to present the (hi)story of Corna. After that, I will describe this (hi)story composed of *'nea* M.'s account (in which I recognised facts, happenings and ways of narration, which emerged in some form from interviews with other conversation partners as well) and my own analysis (which is informed by stories and bits of information presented to me by all my other interlocutors from and connected to the area). Finally, I will conclude my chapter by pointing to how a somatic poetry of the young population from Corna might grow into different memories of the past in the way they advance into their future differently. This emphasises the awkwardness of telling (hi)stories in the process of living them.

A Context for *Now*

The landscape of Corna is one typical to the Apuseni Mountains of the Romanian Carpathians, a group of mountains with its highest peaks ranging between 1400 metres and 1800 metres above sea level. The villages in these mountains are relatively densely populated, the houses have small gardens and are close to each other. Most of the agricultural work is done on the margins of the villages and consists of harvesting hay for the few cattle that locals keep for personal use. Corna itself is situated in a valley with steep slopes surrounding the two sides of the road that slithers down, connecting the village centre with the town of Abrud. The houses line up (or used to line up) next to each other on the two sides of this road, only occasionally forming clusters of two to five houses further up or deeper in-between the hills.

The Apuseni are particularly rich in deposits of iron, gold, silver, copper, coal and other precious or rare minerals. This is important, because what lies in the depths has defined the activities of those living on the surface for decades or even centuries, and continues to have a strong hold on communities even where mining activities have ceased. (In fact, in many places where the extraction of natural resources has stopped, the mining of cultural and social resources has started. Enter anthropologists, sociologists, historians, politicians etc.: new types of exploiters, who instead of mining the depths of mine galleries dive into the intricacies of human minds and relationships!).

The village of Corna was proposed as the site of the tailing pond for the new Roșia Montană mine more than twenty years ago. In 1996 a Canadian company, Gabriel Resources Ltd. came with a proposal to build Europe's largest open-pit goldmine on the site. They obtained a twenty-year lease from the government for mining purposes (see also Alexandrescu, 2012) and proposed to extract more than 300 tonnes of gold and around 1600 tonnes of silver from the mountains Cârnic, Cetate and Jig-Văidoaia (Timonea, 2015). The Roșia Montană site of the state-owned Minvest Deva S. A. gradually closed down from the 1990's, completely ceasing mining activities in 2006, and making 427 employees redundant at the time of its closure (Timonea, 2006; see also Egresi, 2011). For the new project to start, however, most of the population of the village of Roșia Montană, and all of the population of Corna, would have to be resettled. From 2002—despite not having obtained the necessary permits to start mining even to this day—the company started buying people's lands and properties, using various techniques to pressure or convince them (Velicu, 2014; Velicu and Kaika, 2017; Vesalon and Cretan, 2013). As a result, more than seventy percent of the population of the two villages have moved away between 2002 and 2007—something the remaining locals refer to as "the great exodus"—and more than eighty percent of the lands in and surrounding the villages of Roșia Montană and Corna belong to Roșia Montană Gold Corporation (RMGC). RMGC was founded in 1997 as a joint venture of Gabriel Resources Ltd. (80.69%) and the Romanian state (19.31%).

Although most (if not all) of the local population was opposed to the new mine at the beginning, the way unemployment rose, and the ways economic and social insecurity seeped into the village, prompted many people to reconsider their position *vis-a-vis* the company. In 2002 the county council declared Roșia Montană a mono-industrial mining area (Ivanciuc, 2013), cutting access to European funds, and halting all other types of development. In the late 2000's and early 2010's RMGC was the only opportunity for employment within the commune. The arguments of older local miners and Romanian environmental and economic experts,[5] who considered the project unfeasible from socio-economic and ecological perspectives, and unsustainable in the long run, faded into the background from the perspective of locals, who experienced dramatically worsening economic conditions.

As an initial response to the mining project, in 2000 locals (aided by Romanian and international activists) formed the Alburnus Maior Association, whose primary aim was to promote sustainable alternatives to mining, while opposing the Canadian project in an organised manner. Today, however, many of the people still living in the village find it difficult to identify with this agenda, as they find the discourse of the opposition—which became increasingly a narrative about the necessity to protect the environment, archaeological, and mining heritage—less and less relevant to their empirical realities, which were more readily defined by economic and social lack. Anti-mining activists present tourism and agro-tourism as the most viable economic alternatives, but they often fail to tackle the socioeconomic disparities that prevent a large part of the local population from investing in such businesses. In an attempt to achieve permanent protection against large-scale industrial developments, NGOs (including the Alburnus Maior Association) started advocating for the inclusion of Roșia Montană as one of the UNESCO World Heritage sites. The inclusion amongst the

5 The Romanian Academy of Sciences was firmly against the project from the beginning. In their official statement regarding the project they express concerns about the negative environmental, economic, social, and cultural impacts that the new mine would have. See: https://acad.ro/rosia_montana/pag_rm04_decl.htm.

UNESCO sites would somewhat protect Corna from the environmental consequences of the unsustainable mining project, but would also limit the possibilities for development in the area, where (heavy) industry has been the main economic driving force for decades. This leaves many locals torn between what seem to be the only two equally unappealing alternatives, neither of which has any connection or regard to the ways they understand and experience their relationship to each other, to the landscape and to 'Others', or to the ways they perceive their identity, which is rooted in their stories of how they have 'traditionally' lived through changing historical times.

A (*Hi*)*Story* of Dwelling in Corna, Roșia Montană

Roșia Montană, dating back to 131AD, is the first documented settlement in Romania. An important (gold)mining and trade hub belonging to the Roman Dacia Superior, mining was the primary occupation in Alburnus Maior (the Roman name) since time immemorial (Ion, 2014). Their identity as miners is stressed not only by the memory of the fifty years of socialist industrial revolution in the area, but is also evidenced by the landscape of archaeological findings of Roman and pre-Roman-time mining galleries and cliffs (ibid.). Furthermore, there is a complex web of family histories that connects nearly every local person to mine openings, through a deep sense of (lost) ownership, as before collectivisation most families either owned or rented a mine entrance. This was a necessary step for being able to make a living in Corna and Roșia Montană in the first half of the twentieth century.

An ideal picture of the village lives in the memories of the local people, and that picture invariably dates back to this era before communism. It is a frozen instance in time that depicts everyday, hard, and often unrewarded work—that men, women and children took their share of—as a quasi-fairytale ideal. '*Nea* M. (63) recounts:

> So, this has always been a mining community, one hundred percent. In what sense? They have mined gold here for centuries. There were privately owned mines until 1948. Every family had a stamp mill, which used the power of water to break stones. Do you understand? Now, they used these installations only when the waters were not frozen over. They were not operational during the winter. This was the livelihood of people. [...] So the men worked, they carried the ore with wagons

> from the inside of the mountain, from where the women and children transported the stones with horses to the stamp mills. There, the head-sized rocks were broken into smaller pieces (first of approximately the size of a fist, and later, at a different installation these were further smashed into powder, until the gold, silver and pyrite could be separated on a woolen reel. They then washed this sand from the reel and gathered the gold powder into a bigger piece with mercury. This they could then sell). This is how people around here lived up until '48.

There is a rhythm to this fragment of narrative through which *'nea* M. describes the pulse of life in a time that he himself only knows from the memories of an older generation, a generation that is no longer able or willing to tell their stories. It is one that evokes the cyclical time 'remembered' from the everyday of the pre-socialist era, a time defined for the people of Corna by the cycle of finding–working–selling of gold. This was a time when making a living involved direct, often intimate or dangerous interaction with the ore by all (or most) members of a family, a time when locals depended on the qualities of the landscape emergent in all their dynamic interactions. This was, for example, a time when it was not uncommon for men, women and even children to encounter the *vâlva băii,* the elusive fairy of the mines, who held all knowledge of the whereabouts of gold veins and had the power to share it, but who also caused the death of many miners by leading them astray.

Although the hardships, unpredictability and dangers of the lives of miners is often stressed when villagers talk about this time, they also attach a beauty to this rhythm of life. Almost all my interlocutors talked about beauty with a sense of loss, whether they mentioned the aesthetic qualities of landscape, of the culture or of the lively social life, and many people talked about the beauty of their distinct way of life in a time that pre-dated their own lived experiences. In *'nea* M.'s account the first instance that emphasised a change from the way their life went "as usual" was the instalment of the socialist state and the socio-economic changes that resulted:

> Then, in 1948 nationalisation happened. They closed all the stamp mills... they took away the livelihood source of people from these villages! Those communists came and they [...] carried the miners away to work, to Salva Vișeului and other places. To work on building sites, the poor men! And the women remained here alone, with the children. They [the communists] said that they were rich, they should hand over their gold.

> But my grandmother told me, that there had been months when they could barely afford 15 kg of cornflour. [...] Then they opened the state mine and employed the men there. It was in the fifties and sixties when the mine really began to work well, but they were working underground and many got ill from silicosis. They were called Stakhanovite, their photographs were put on billboards, they received awards, money for their hard work! But by the time they managed to build a house, put a roof on it, they died, leaving their wives and children behind. Then they opened the pit in the seventies.

There is a duality in the way people from Corna remember the socialist period. On the one hand, before the late seventies working in mines had very grave health consequences and nearly all of my interlocutors had at least one male family member (father, husband) who had died as a result of working in the mine. On the other hand, however, as *'nea* M.'s account also reveals, there was also a sense of growth in terms of personal and collective economic conditions and an overall improvement of the status of the mining class (see also Kideckel, 2004; Vasi, 2004).

One can notice the change between the two depicted time periods even in the narrative, ethnopoetic style as the pulse becomes somehow suppressed, and a series of events are recounted in a successive manner. Although working in the goldmine as a state employee had its own rhythm in the everyday toil of being a miner (at least for adult male members of the family), and the agricultural cycles continued to define a large chunk of the family's life, the focus of the story here shifts to plans and outcomes, which seems to parallel the socialist pursuit of development and modernity (see also Verdery, 1996: 39–57). Life is no longer made up of the rhythm of gold finding–working–selling cycles, but adopts a more linear understanding, where the emphasis is on achieving life goals. Time becomes flattened in the sense that life*lines* no longer reflect the rhythm of interpersonal and ecological relationships, but rather, they become a connecting dash between a starting and an ending point. In this type of story there is no longer an intimate connection between miner and ore. Gold (and implicitly the landscape) as a means of sustenance and the ecological relationships, that defined everyday living, lose their central roles in favour of the mine as a *state*-owned enterprise, which provides its workers with everything that was previously connected to the soil and the mineral: livelihood, relationships, status.

An important aspect of the post-socialist experience of the working class is that many people observed the greatest technological advancements in industry and entertainment during the socialist period, and these innovations had a positive impact both on their social and on their cultural lives during that time (see also Hann et al., 2002). There was a cinema, a casino-ballroom, food markets, social and cultural events in Roșia Montană during the communist reign, all of which were closed down during the 1990s. Many are nostalgic when remembering the miners' choir, and the fact that the state-mine had a football team. Moreover, during the socialist time workers in heavy industries were highly regarded, and all (post)industrial communities in eastern Europe are greatly defined by this period that they experienced and now perceive in terms of stability (Wódz and Gnieciak, 2017), clear life trajectories (Pine, 2017), and a sense of worth (Kalb, 2017; Pine, 2017; Wódz and Gnieciak, 2017).

These elements are all experienced and incorporated into post-socialist realities and life stories in terms of loss. Miners were a privileged group of workers, enjoyed many benefits, had a very strong union, and a complex social network that extended beyond their workplace through the activities of their wives and families (Kideckel, 2002; 2004; Vasi, 2004). The growing class differentiation in post-socialist Romania, however, left the working class—especially those (formerly) employed in the heavy industries—in a disadvantaged position. Former miners from Roșia Montană talk about the social and economic security from before the fall of the communist regime and always juxtapose it with the current precarity of young people, the high rates of unemployment and general social and economic insecurities (Velicu, 2012a; 2012b; see also Narotzky, 2017).

In arriving in the present day in his account, the sudden change in style of *'nea* M.'s monologue also expresses these insecurities and the frustrations of the community, which is unable to reach a resolution to the current situation:

> That is when [in the '70s] they gave up those ideas, that these rocks are valuable and we should protect them. Now everyone thinks, that wow, we should marvel at the nice rocks and the mining landscape in Rosia Montana and die of hunger! There is an unemployment rate of 80–90%, we don't have anything to live from and now they want to include it in

> the UNESCO!... So that you can live off a little cow?! You cannot live off that! You can only do semi-subsistence farming here! On a small scale, not like on the plains!... And the young people no longer stay here! [...] Okay, if they don't want the Canadians to take the gold, the Romanian government should invest in it. Why? Is this place so beautiful that we cannot touch it?! [...]

The shift from socialist stability to this state of precarity, however, was not as abrupt as the swift cut in '*nea* M.'s narrative flow would suggest. The events leading to the closure of the state mine were a gradual process, after which the company employed a strategy of slowly 'easing the people into' the new situation. Throughout the 2000s, RMGC slowly took over the position of 'patron' in Roșia Montană—previously the role of the state—by providing jobs (which were temporary), by filling in some of the missing infrastructure, such as bus routes to the nearby towns of Câmpeni and Abrud (which were not in any way sufficient to meet the needs of the local population), and even by providing social assistance to some of the locals (who were 'forgotten' by state institutions). The much less mediatised village of Corna was, however, almost entirely neglected as far as infrastructure was concerned and locals became more and more isolated both from each other and from other communities.

Most of the adult population from both villages have worked for the company at some point between 2001 and 2015, when in the increasingly acute unemployment crisis after the gradual closure of the state mine, *Goldul* (RMGC) was the only employment opportunity in the area. This 'opportunity', however, only increased insecurity, as envy and suspicion rose amongst locals, and it is rumored that people were pressured into selling their lands and houses with the promise of a job (see also Velicu and Kaika, 2017).

Locals refer to the company as *Goldul* (the Gold). This reveals an interesting process through which the social, symbolic and economic values attached to the gold mineral (a source for the livelihoods of locals for millennia) was transferred to Gold, the company which in the post-socialist liquid modernity first assumed the socio-economic (and even cultural) role that the state(-mine) had during socialism, and later became the denominator of value and the only potential source of livelihood, development and future for the youth in Roșia Montană.

In the socio-economic equation of the taskscape of Roșia Montană, thus, gold(mining) in pre-socialism was what (state) mining was in socialism, and is what the Gold (Corporation) is in post-socialism. *Aur* (the Romanian word for gold), however, became something strange in this process, alienated from them by the foreign company and the foreign word. Whereas there were heightened expectations attached to the technological and economic changes a transition to free-market economy would bring, their experience of the post-socialist capitalism of the late 1990s was simply one of loss and disempowerment as economic and social resources grew more and more out of their immediate and symbolic control. This estrangement was felt in changes in everyday activities, but also quite literally in the 2000s, through losing ownership and control of surrounding lands.

Stories Unfolding

A total of forty-two households remain in Corna today, within which there is usually only one elderly person or an elderly couple. Most people have as neighbours long stretches of uninhabited lots: the empty spaces of houses that were demolished shortly after being bought by the RMGC. Demolition has left locals in a vulnerable position and with visible signs of their disrupted social fabric. In the dynamic interaction of dwelling, in which people and the landscape are in constant co-constitution, broken or waning social relationships also mean dissipating ecological connections. The forest has almost entirely recovered from its previous state of over-harvesting, and its foliage now covers the former hayfields and has started to grow on the empty house lots, inviting wild boar, foxes and other animals into these places, into the landscapes where the human involvement with other human and non-human actors, including the activities of making homes and making a living, were much more visible. Thus, the landscape, this ongoing conversation between people and their environment (Ingold, 2000), has somewhat lost part of the human 'voices' that were so intrinsically a part of Corna.

The temporality of the landscape becomes questioned along with the disappearance of "histories"—which in our case includes, but is not limited to, the dynamic interchange between man and minerals—"that are woven, along with the life-cycles of plants and animals into the texture of the surface itself" (Ingold, 2000: 198). Many of my interlocutors lamented the fact that it is no longer possible for them to walk on the road after dusk: the street lighting hardly exists anymore, the distances between inhabited houses are far too big, and the roads have become populated by wild animals. The once dynamic network of mutually affecting correspondence between different types of actors (Ingold, 2017) has dramatically changed, and both human and ecological relationships have become uneasy.

The ongoing dispute and the resulting disruptions in relationships and in personal and community narratives create a state of permanent liminality (Horváth, 2013; Szakolczai, 2015a; 2015b; Thomassen, 2015; Wydra, 2000; Wydra et al., 2015; see also Sampson, 2002), where the people of Corna experience everything as fragmented, fragile, inconsistent, and provisory. Although unpredictability and insecurity are experiences that resurface time and again when narrating the (hi) story of Corna, the current empirical realities connect to a wider context of similarly destabilised post-industrial and post-socialist places and communities (see Bridger and Pines, 1998; Narotzky and Goddard, 2017). Whereas the accounts of the pre-socialist time present difficult periods, inherently vulnerable and unpredictable cycles in the acquiring of what was necessary for oneself and for the family, this was something immanent to the *certitude* of an ongoing relationship with the landscape, of a life lived in a mining environment (which in socialism transformed into the stability of employment in the state mine). The fragility of their livelihood today, however, is on the one hand evidence of disrupted (social, ecological, economic and intergenerational) relationships (see also Pine, 2017; Kalb, 2017), and on the other hand, an acute experience on the micro level of the present erratic macro-economic and state-political processes (Goddard, 2017; Wódz and Gnieciak, 2017). This latter aspect determines their long-term and everyday economic decisions, as many become uninterested or afraid to invest and plan for the future in a context where they see everything as dependent on a single decision that keeps getting postponed.

The disrupted social and ecological networks create a background where the young population from Corna (the young adult children of those who did not relocate during the "great exodus") find themselves in a peculiar position, especially if their understanding of themselves and their future prospects incorporates feelings of belonging to the place. The (hi)story of dwelling in Roșia Montană is a complex one that should include the multiple voices, which—through the principle of human correspondence (Ingold, 2017)—co-create each other. The somatic poetry of the youth, however, is entirely different from that of the older generation in the ways that it stresses beauty and belonging differently, in the calm of nature and feelings of homeliness *vis-à-vis* the bustle and foreignness of the city or of a different country. Young people from Corna today are either unemployed and/or do seasonal work abroad, or have migrated to cities, where they either have stable employment or change their jobs frequently, returning to Corna occasionally for periods between two jobs or for short visits 'home' to rest. Only some of the extremely few, who have family members employed at the copper mine at Roșia Poieni (the only ones in the area who still work in a mine) are hopeful that they will manage to 'get in' and secure a job there as well—a form of the reproduction of the working class not uncommon to industrial and post-industrial communities (see Perelman and Vargas, 2017).[6] In general, however, they seem uninterested in mining and do not consider working in the mine as a career option.

A continuation of the mining (hi)story, thus, is revealed as the vision of the older generations, rather than a necessity for the younger ones. Neither the personal and collective memories or stories, nor the aspirations of the young adult population of Corna are connected to the goldmine. While—similarly to their elders—they do understand their (hi)story in terms of loss and they perceive their future prospects to be thwarted because of how the community changed as a result of the activities of the RMGC in the area, they do not necessarily perceive

6 The copper mine at Roșia Poieni is also on the territory of the Roșia Montană commune, but is operated by a different state-owned company, *Cupru Min S. A.* It is Romania's largest, and Europe's second largest, copper reserve. The mine is open-pit and the tailing pond was created by flooding the village Geamăna. The polluted lake with the still visible bell-tower is often photographed by media and activists to illustrate the effects mining pollution can have on the environment and on communities. Comparison is often made between Corna and Geamăna due to their possibly analogous history. However, the apparent demise of Corna seems to be going in a very different direction: instead of socialism, it is a post-socialist capitalism that forces people to leave, and instead of being filled (with polluted waste) and taken over by industry, it is emptied (of built history) and taken over by the regenerating force of an increasingly hostile and unfamiliar natural environment.

the discontinuation of gold-mining as the most important aspect of these changes, and they are very rarely involved in the discussions regarding the protection of the heritage of Roșia Montană. They are in-between (places, memories, future possibilities), and they navigate the unpredictability of the post-socialist, post-industrial condition, highlighting only the loss of reliable connections (Narotzky, 2017).

I have suggested before that life and story are inseparable in the (hi) story of dwelling, a (hi)story that imbues the landscape with aesthetic qualities that unfold into the individual and collective somatic poetries, into a sense of beauty and belonging. Whereas the experiences of loss of this younger generation connect to different aspects of life and memory, their understanding of beauty in dwelling also unfolds in response to conversations with a landscape and with all types of different actors, who are connected in a variety of ways. Although at this point their stories remain untold, both due to a still heightened interest in what is perceived as the 'real' (i.e. concluded) history of the place, and due to the difficulty of narrating life in the process of living, I believe that these new poetries will undoubtedly affect the ways the (hi)story of dwelling in Corna might develop in the coming years.

References

Alexandrescu, F. M., 2012. Human Agency in the Interstices of Structure: Choice and Contingency in the Conflict over Rosia Montana. University of Toronto (Unpublished doctoral thesis). http://hdl.handle.net/1807/32297.

Bauman, Z., 2000[1999]. *Liquid Modernity.* Polity, Cambridge.

Bridger, S., and Pine, F. (Eds.), 2001. *Community. Seeking Safety in an insecure world.* Polity, Cambridge.

——, 1998. *Surviving Post-Socialism: Local Strategies and Regional Response in Eastern Europe and the Former Soviet Union.* Routledge, London and New York. http://doi.org/10.4324/9780203350669.

Egresi, I., 2011. The Curse of the Gold: Discourses Surrounding the Project of the Largest Pit-Mine in Europe. *Human Geographies — Journal of Studies and Research in Human Geography*, 5 (2), 57–68. http://doi.org/10.5719/hgeo.2011.52.57.

Eriksen, T. H., 2016. *Overheating: An Anthropology of Accelerated Change.* Pluto Press, London. http://doi.org/10.2307/j.ctt1cc2mxj.

Goddard, V., 2017. Work and Livelihoods: An Introduction, in: Narotzky, S., and Goddard, V. (Eds.), *Work and Livelihoods: History, Ethnography and Models in Times of Crisis.* Routledge, London, pp. 1–27. http://doi.org/10.4324/9781315747804.

Hann, C., Humphrey, C., and Verdery. C., 2002. Introduction: postsocialism as a topic of anthropological investigation, in: Hann, C. (Ed.), *Postsocialism. Ideals, Ideologies and Practices in Eurasia.*: Routledge, London and New York, pp. 1–28. http://doi.org/10.4324/9780203428115.

Horvath, A., 2013. *Modernism and Charisma.* Palgrave Macmillan, London. http://doi.org/10.1057/9781137277862.

Ingold, T., 2017. On Human Correspondence. *Journal of the Royal Anthropological Institute*, 23(1), 9–27. http://doi.org/10.1111/1467-9655.12541.

——, 2000. *The Perception of the Environment. Essays in livelihood, dwelling and skill.* Routledge, London. http://doi.org/10.4324/9780203466025.

Ingold, T., and Hallam, E., 2007. Creativity and Cultural Improvisation: An Introduction, in: Ingold, T., and Hallam, E. (Eds.), *Creativity and Cultural Improvisation.* Berg, Oxford and New York, pp. 1–24. http://doi.org/10.4324/978100313553.

Ingold, T., Lowenthal, T., Feeley-Harnik, G., Harvey, P., and Küchler, S., 1996. 1992 Debate: The Past is a Foreign Country, in: Ingold, T. (Ed.), *Key Debates in Anthropology.* Routledge, London and New York, pp. 161–84. http://doi.org/10.4324/9780203450956-11.

Ion, A., 2014. Roșia Montană: When Heritage Meets Social Activism, Politics and Community Identity. *Online Journal in Public Archeology*, 4/2014, 51–60. http://doi.org/10.23914/ap.v4i0.43.

Ivanciuc, T., 2013. De ce nu merge turismul la Roșia Montană [Why tourism doesn't work in Roșia Montană]. http://www.contributors.ro/economie/de-ce-nu-merge-turismul-la-rosia-montana/.

Kalb, D., 2017. Regimes of Value and Worthlessness: How Two Subaltern Stories Speak, in: Narotzky, S., and Goddard, V. (Eds.), *Work and Livelihoods: History, Ethnography and Models in Times of Crisis.* Routledge, London, pp. 123–36. http://doi.org/10.4324/9781315747804-17.

Kideckel, D. A., 2004. Miners and Wives in Romania's Jiu Valley: Perspectives on Postsocialist Class, Gender, and Social Change. *Identities: Global Studies in Culture and Power*, 11(1), 39–63. http://doi.org/10.1080/725289024.

——, 2002. The Unmaking of an East-Central European Working Class, in: Hann, C. (Ed.), *Postsocialism. Ideals, Ideologies and Practices in Eurasia.* Routledge, New York, pp. 114–32. http://doi.org/10.4324/9780203428115-11.

Narotzky, S., and Goddard, V. (Eds.), 2017. *Work and Livelihoods: History, Ethnography and Models in Times of Crisis.* Routledge, London. http://doi.org/10.4324/9781315747804.

Sampson, S., 2002. Beyond Transition: Rethinking Elite Configurations in the Balkans, in: Hann, C. (Ed.), *Postsocialism. Ideals, Ideologies and Practices in Eurasia.* Routledge, London and New York, pp. 297–316. http://doi.org/10.4324/9780203428115-23.

Szakolczai, A., 2015a. Liminality and Experience: Structuring Transitory Situations and Transformative Events, in: Horvath, A., Thomassen, B., Wydra, H. (Eds.), *Breaking Boundaries: Varieties of Liminality.* Berghahn, Oxford, pp. 11–38. http://doi.org/10.2307/j.ctt9qcxbg.5.

——, A., 2015b. Marginalitás és liminalitás. Státuszon kívüli helyzetek és átértékelésük. *Regio*, 23(2), 6–29. http://doi.org/10.17355/rkkpt.v23i2.48.

Tarkowska, E., 2006. Zygmunt Bauman on Time and Detemporalisation Processes. *Polish Sociological Review*, 3(155), 357–74.

Thomassen, B., 2015. Thinking with Liminality: To the Boundaries of an Anthropological Concept, in: Horvath, A., Thomassen, B., Wydra, H. (Eds.), *Breaking Boundaries: Varieties of Liminality.* Berghahn, Oxford, pp. 39–58. http://doi.org/10.2307/j.ctt9qcxbg.6.

Țimonea, D., 2015. Moştenirea din Apuseni: cât aur era acum un veac şi cât a mai rămas, de fapt, la Roşia Montană [The legacy of Rosia Montana: how much gold was there a century ago and how much is there left]. https://adevarul.ro/locale/alba-iulia/mostenirea-apuseni-aur-era-veac-mai-ramas-fapt-rosia-montana-1_54df19f5448e03c0fd993205/index.html.

——, 2006. Exploatarea minieră "Roșiamin" se închide în două săptămâni. ["Roșiamin" exploitation site closes in two weeks]. https://romanialibera.ro/actualitate/proiecte-locale/exploatarea-miniera--rosiamin--se-inchide-in-doua-saptamani--46833.

Uzendoski, M., 2008. Somatic Poetry in Amazonian Ecuador. *Anthropology and Humanism*, 33(1–2), 12–29. http://doi.org/10.1111/j.1548-1409.2008.00002.x.

Vasi, I. B., 2004. The Fist of the Working Class: The Social Movements of Jiu Valley Miners in Post-Socialist Romania. *East European Politics and Societes*, 18(1), 132–57. http://doi.org/10.1177/0888325403258290.

Velicu, I., 2015. Demonizing the Sensible and the 'Revolution of Our Generation' in Rosia Montana, *Globalizations*, 12(6), 846–58. http://doi.org/10.1080/14747731.2015.1100858.

——, 2012a. The Aesthetic Post-Communist Subject and the *Differend* of Rosia Montana. *Studies in Social Justice*, 6(1), 125–41. http://doi.org/10.26522/ssj.v6i1.1072.

——, 2012b. To Sell or Not to Sell: Landscapes of Resistance to Neoliberal Globalization in Transylvania. *Globalizations*, 9(2), pp. 307–21. http://doi.org/10.1080/14747731.2012.658253.

——, 2014. Moral versus Commercial Economies: Transylvanian Stories. *New Political Science*, 36(2), 1–19. http://doi.org/10.1080/07393148.2014.883804.

Velicu, I., and Kaika, M., 2017. Undoing environmental justice: Re-imagining equality in the Rosia Montana anti-mining movement. *Geoforum*, 84, 305–15. http://doi.org/10.1016/j.geoforum.2015.10.012.

Verdery, K., 1996. *What Was Socialism and What Comes Next?* Princeton University Press, New Jersey, NJ. http://doi.org/10.1515/9781400821990.

Vesalon, L., and Creţan, R., 2013. 'Cyanide kills!' Environmental movements and the construction of environmental risk at Roşia Montană, Romania. *Area*, 45, 443–51. http://doi.org/10.1111/area.12049.

Wódz, K., and Gnieciak, M., 2017. Post-industrial Landscape: Space and Place in the Personal Experiences of Residents of the Former Working-class Estate of Ksawera in Będzin, in: Narotzky, S., and Goddard, V. (Eds.), *Work and Livelihoods: History, Ethnography and Models in Times of Crisis.* Routledge, London, pp. 137–53. http://doi.org/10.4324/9781315747804.

Wydra, H., 2000. *Continuities in Poland's Permanent Transition*. Macmillan London. http://doi.org/10.1057/9780333983003.

Wydra, H., Thomassen, B., and Horvath, A., 2015. Introduction, in: Horvath, A., Thomassen, B., Wydra, H. (Eds.), *Breaking Boundaries: Varieties of Liminality.* Berghahn, Oxford, pp. 1–8. http://doi.org/10.2307/j.ctt9qcxbg.4.

PART III

8. The Shifting Geopolitical Ecologies of Wild Nature Conservation in Romania

George Iordăchescu

Wilderness as Political Ecology

Recent debates about biodiversity conservation and habitat protection in Europe—from state governments and Brussels—favour a turn towards strict protection, wilderness frontiers and untouched nature narratives. These raise serious concerns about social and environmental justice. Although there is no clear consensus on defining wilderness for policy-making, many initiatives converge around this aim. Many of these proposals have found fertile ground for experimentation and development in eastern Europe. This chapter explores how newly discovered appreciation for wilderness is set to transform nature conservation in this region by reaffirming older forms of economic dependency and unequal environmental exchange. While zooming in and out on such transformations happening in Romania, the state/conservation nexus is used as a lens to understand the creation of 'Eastern Europe' as a green internal periphery. This chapter will scrutinise the 'Eastern European wilderness momentum' by fleshing out the ongoing creation of a private wilderness protected area in the Southern Carpathian Mountains.

Over the last decade, various MEPs and officials from the European Commission have worked together to advance the protection of wilderness in the Union, issuing soft laws such as the Guidelines for the Natura 2000 protected areas (European Commission, 2013)

 https://doi.org/10.11647/OBP.0244.08

and a dedicated resolution (European Parliament, 2009). In parallel, prominent environmental NGOs initiated concrete actions to identify the last areas of 'unspoiled' nature, to lobby for their strict protection as part of domestic legislation and to turn wilderness conservation into a profitable business through its commodification within ecotourism operations and as part of climate change mitigation strategies. A new re-valuation of old-growth forests and other intact eastern European landscapes have made the region a prime focus for new financial mechanisms for carbon sequestration, biodiversity conservation strategies and new economic growth models (European Commission, 2020).

These new developments target large areas of the eastern EU member states. However, surprisingly, 'wilderness' is not mentioned in any of the national legal frameworks in the region. Rewilding Europe, the Endangered Landscape Programme, EuroNatur and other important civil society conservation actors have concentrated much of their efforts around supporting local initiatives celebrating 'wild' nature, or have started top-down wilderness conservation projects. At the political level, some states have championed this approach to conservation from its infancy (e.g. the Czech Republic as discussed in Petrova, 2013), while others have been somewhat reluctant to value their natural heritage as 'untouched nature' (e.g. Poland as discussed in Blavascunas and Konczal, 2019; and Gzeszczak and Karolewski, 2017). As civic campaigns and high-end political negotiations around wilderness protection turn the issue into a recurrent topic on the public agenda, the geopolitical nature of this conservation approach becomes more critical in redefining the ways borders and peripheries are understood and acted upon in the region (Wild Europe, 2019). Although very heterogeneous, all these projects and initiatives share a few standard features: they come as a response to degradation narratives or land abandonment, and propose wilderness conservation as a way to fix these problems; they present a strict protection approach opposed to an allegedly failing marginal agriculture; they legitimate the interventions by appealing to Western scientific knowledge; and lastly, they glorify past ecological riches that western Europe has lost, augmenting the urgency to act. Through a political ecology approach, this chapter discusses power, knowledge production, environmental justice and hegemonic conservation narratives associated with the re-valuation of wild nature in eastern Europe with a focus on Romania.

This chapter does several things. First, it shows that wilderness protection is gaining momentum in eastern Europe and that this process enjoys the blessing of various governments. Second, it details this relation by scrutinising an ongoing establishment of a private wilderness reserve in the Southern Carpathians in Romania as well as the negotiation of a legal frame for the strict protection of 'virgin' forests by a technocratic government. Finally, it shows that wilderness conservation in Romania reinforces unjust dependencies and new forms of accumulation as wild nature becomes an environmental fix.

Conservation Geopolitics

Emerging from civil society struggles or as private projects, wilderness-related enterprises have been championed by state and regional authorities throughout the entire eastern European region. As a new Common Agricultural Policy and a European Green New Deal are implemented, conservationists have suggested that a growing interest in conserving 'untouched' nature will mark a new era in intergovernmental cooperation and will conclude with the introduction of 'wilderness' values in sectors such as agriculture, energy and infrastructural developments (Wild Europe, 2019). Intensely lobbied for by a coalition of environmental NGOs, scientists and philanthropists, wilderness debuted on the EU political scene with the adoption of a resolution by the European Parliament on 3 February 2009 (European Parliament, 2009).

'Wilderness' as a concept of concern for environmental law and policy-making in the EU is very young (Egerer et al., 2016). The current wilderness momentum needs to be historicised and investigated against contemporary global conservation debates. I join others in reconsidering the regional specificities of wilderness preservation in the European context (Lupp et al., 2011; Lupp et al., 2012; Kupper, 2014; Kirchoff and Vicenzotti, 2014). I argue that the local historical and socio-political context makes eastern European wilderness protection significantly different to other, similar movements. Far from adopting a globalised, Yellowstone fortress-type of narrative, European actors propose many interpretations of wilderness, each with profound political and social implications (Saarinen, 2015; Bastmeijer, 2016; Schumacher, 2018). While public attitudes to wilderness vary (Bauer et al., 2017), most of the recent legal developments champion a strict separation of wild nature from human history and use (Martin et al., 2008; Wild10, 2015; Egerer et al., 2016).

Read as part of a global attempt to strictly secure large areas of land for nature to develop according to its own rules, the European wilderness momentum appears as a process of re-territorialisation on the one hand (Adams et al., 2014) and as the creation of a new resource on the other. The new resource has become of utmost importance amongst EU strategies for green growth and climate change mitigation. The making and maintenance of these resources have involved the establishment of strict boundaries between domesticated nature and the areas in which (mostly white male) scientists and conservationists 'discovered' an autonomous nature that has evolved independently of any human influence. These boundary-drawing dynamics will be investigated through the Carpathia Project in the Southern Carpathian Mountains. The project under scrutiny changed not only local socio-environmental relations, but also the wider political economy of the area.

There is one particular process of capitalist transformation of nature into a commercial value that is more prevalent than others in the creation of the eastern European wilderness frontier. This is the 'cheapening of nature', a process of control and devaluation of nature as a source of essential inputs for the development of global capitalism (Moore, 2015; Moore and Patel, 2018). Adapted to local realities, the 'cheapening of wilderness' is a foundational moment for strict conservation initiatives in eastern Europe. This process is intimately imbricated within recent historical events such as land restitution and reform, the devaluation of the forest by illegal logging and deforestation, the top-down establishment of protected areas and a constant depreciation of traditional livelihoods.

The Romanian forests of the Făgăraș Mountains are heavily impacted by extractivist processes and are considered to be of particular ecological value by non-state conservation programs. The Carpathia Project is legitimised by its promoters as undoing some of the environmental harm done by recent ruthless timber exploitation. While stopping commercial logging and hunting, the Foundation Conservation Carpathia aims and succeeds to buy as much land as there is available, claiming that exclusive (private) ownership is the safest strategy for strict protection in perpetuity.

This case study is informed by interviews with people working for the implementing organisation, direct observation, field visits and the study of legal documents, grey literature, technical reports, wildlife documentaries and several other media productions. As the project is situated within a highly political field of negotiating new values attached to nature, I follow the Foundation Conservation Carpathia (FCC) as an actor involved in building the first eastern European private wilderness reserve.

Eastern Europe as a Wilderness Frontier

Over the last ten years, the protection of 'wild nature' has gained increasing momentum in Europe. From the extensive mapping of remaining wilderness to progress with EU legislation, proposals for the strict protection of nature have set the ground for many continent-wide alliances and permeated national and institutional boundaries. Although merely a decade old, such conservation approaches have triggered important changes in socio-environmental relations. I argue that these wilderness protection projects have predominantly targeted the outer regions of the EU, creating an imagined green periphery. As I am focusing on such processes developing in eastern Europe, I propose to call this green internal periphery 'The New Wild East'.

The New Wild East represents a politico-environmental frontier whose importance goes beyond nature protection and is underlined by spiritual values, productive aesthetics and a lot of experimentation. As it is read from the 'West', this wilderness frontier was revealed and subsequently discovered after the fall of the iron curtain. The official storyline goes like this:

> the fall of the iron curtain, [...] revealed large, intact areas in central and Eastern Europe, primarily along the east-west border, and created significant opportunities for government-protected areas (Martin, 2008: 34)

Since the wild nature of eastern Europe and the wilderness in the periphery have been 'discovered', threatening degradation narratives have proliferated. Overgrazing, intensive use, deforestation, overhunting, highway and infrastructure development, are all elements of a sudden attack on Europe's last wild areas. For example, damming in the Balkans is destroying Europe's "blue heart" (EuroNatur, 2016), illegal logging is a threat to the last "remaining wilderness" of Poland (Gross, 2016). On the other hand, these threats are rapidly turned into opportunities for conservation:

> Conservation organisations today have the unique opportunity to acquire large areas of land to secure in perpetuity. Ecological and evolutionary processes can be allowed to convert landscapes that still possess wilderness qualities and ecological richness back into true wilderness for the benefit of biodiversity and the people alike. (Promberger, 2015: 242)

If we aim to interpret this eastern European wilderness momentum as a creation of a green internal periphery, it is essential to ask ourselves whose periphery would the New Wild East be relative to? Who are the human and the more-than-human winners and losers of this process? And what can political ecology say about it?

Land abandonment is an opportunity to move towards a new wilderness. In this narrative, the processes underlying land abandonment are unquestioned and rewilding comes as a restorative process "in which formerly cultivated landscapes develop without human control" (Hochtl et al., 2005: 86). Within this new conservation ethic, land abandonment is productive (Jørgensen, 2015: 484), but the underlying causes are always left unaddressed (Tănăsescu, 2017). In eastern European countries land abandonment is often a result of rural under-development, a lack of infrastructure, healthcare, education opportunities and jobs, huge rural-urban investment and livelihood divides and a steady devaluing of agricultural work combined with a lack of outlets for selling the fruits of this work (Fox, 2011).

Closely connected to land abandonment is the issue of rural depopulation. Many wilderness protection projects celebrate so-called wildlife returns across the continent. Leaving aside the fact that only 'charismatic' species seem to return (brown bears, wolves, lynxes), such processes happen predominantly in areas affected by out-migration, ageing populations and other negative demographic trends. From the

Fig. 1. Abandoned land in the Southern Carpathian Mountains. Photograph by George Iordăchescu (2019).

Alpine communities to the Spanish comunales, depopulation seems a critical process negatively affecting environmental stewardship. In eastern Europe, one of the first rewilding projects in the early 1990s was the reintroduction of Konik horses in the Pape region of Latvia, an area marked by massive outmigration, an ageing population and a total absence of markets for local products (Schwartz, 2006). Previous Soviet rule had transformed both the rural economy and the cultural landscape around Runcava village. While pre-Soviet fishing practices were abandoned as the area became militarised, families moved to the bigger cities, leaving the land almost deserted. When a rewilding project started to be considered as feasible, locals still present were (re-)trained to see the land in terms of sustainability, biodiversity and restoration. However, donors chose Pape not only for its ecological riches and sparse population, but also for its low wages and prices. At the same time, the region was close enough to countries like Sweden and Germany, from where potential tourists could come once the new wilderness was established (Schwartz, 2006: 159).

It is important to note that, except for Finland, no EU member country has so far adopted explicit legislation for wilderness protection (Bastmeijer, 2016). Moreover, local grassroots support for wilderness protection in eastern Europe has been weak so far, even if the concept is widely popular in the West (Urban, 2016). Nevertheless, the eastern part of the continent occupies the centre stage for some of the most notable and well-funded strict protection projects, financed by private actors or through public-private partnerships. A quick overview of the European Rewilding Network, a pan-European movement connecting all rewilding initiatives since 2013, shows that twenty-three out of around sixty rewilding initiatives are located in former post-socialist countries (Rewilding Europe, 2020). Moreover, Rewilding Europe, the agenda-setting actor in this field on the continent, has so far established five of their seven rewilding areas in eastern Europe.

Another example is the Endangered Landscape Programme, a recently launched program aimed at supporting remarkable environmental restoration projects for an extended period to secure their success. Financed by the Arcadia Foundation and managed by the Cambridge Conservation Initiative, the programme announced its first round of projects from March 2019. Five out of a total of eight projects receiving support are located in eastern Europe or its immediate vicinity, and their central long-term goal is "to give space back to nature" (Endangered Landscapes Programme, 2017).

EuroNatur, Germany's oldest and most important foundation advocating for wild nature protection, is involved in nineteen projects across the continent, of which thirteen are located in eastern Europe. One of its most ambitious initiatives is the European Green Belt, an initiative aiming to protect and promote the strip of land formerly known as the iron curtain. Stretching over more than 12,000 km, the former demarcation line between east and west is allegedly Europe's "precious natural pearl necklace" consisting of "pristine forests and swamps, wild mountain ranges and river landscapes that cannot be found anywhere else in Europe" (European Green Belt, 2018). In Romania, EuroNatur is one of the leaders and a generous supporter of an environmental campaign for the protection of 'virgin' forests.

These projects attempt to define wilderness uniformly to build scientific coherence and homogenise tools and indicators by assembling pan-European standards, reference indicators and a uniform set of criteria. The strict separation of the newly discovered wild nature from managed landscapes and socio-historical natures is another facet of the same process. This strict separation is necessary and directly impacts on local strategies for rural development, frames imaginations for the future of humans' relations with the environment, and often contradicts locals' aspirations and perspectives (Schwartz, 2006; Petrova, 2013).

This review has tried to demonstrate the apparent abundance of wilderness to be saved in eastern Europe. Since Romania is widely regarded as containing the highest percentage of 'virgin' forests (UNESCO, 2017), charismatic wildlife (Schlingemann et al., 2017; European Parliament, 2018) and 'intact' landscapes, it makes for a good example to study the European shift towards wilderness protection.

State-Conservation Entanglements in Romania

Over the last couple of years, the Romanian government, environmental NGOs, and other actors involved in conservation have actively promoted the country as a biodiversity hotspot and an untouched nature destination. In terms of legislative developments, these efforts have been mirrored by proportional developments that reached a peak while a technocratic government was in office between 2015 and 2016. For almost a decade the country's touristic brand played on narratives of wild nature and adventurous discovery (Iordachescu, 2014), and the government used various diplomatic occasions to portray Romania as the "green heart of Europe" (Romanian Presidency of the EU Council, 2018).

This new valuation of wild nature comes after two decades during which a Carpathian timber frontier has gone from boom to bust (Vasile, 2020), leaving behind an inability to halt illegal logging and deforestation (Iordachescu, 2020). An immediate effect of post-socialist land reforms related to forest privatisation was an increase in timber exploitation. Dorondel describes how both legal and illegal forest exploitation mushroomed within patronage networks (2009), resulting in what he calls "disrupted landscapes" (2016). According to Vasile and others, this post-socialist timber frontier was marked by extensive corruption and violence (Lawrence and Szabo, 2005; Vasile, 2009; Vasile, 2019).

These transformations impacted the region not only from an ecological point of view but also visually. Many forest plots were clear-cut as soon as they were returned to owners. As the timber frontier was coming to an end, wilderness protection became the new hegemonic narrative.

The current grim prospects for nature conservation in Romania were preceded by a series of positive developments under the technocratic government in office between November 2015 and January 2017. That period was marked by signs of progress in laying down the legal framework for identifying and protecting the old-growth forests (referred to as 'virgin forests'), curbing the extent of illegal logging and unwavering support for the creation of wilderness reserves in the Southern Carpathians.

The protection of virgin forests in Romania is a perfect case for understanding regional and even international attempts to conserve wild nature under strict protection regimes. As has been explored above, the abstraction of wilderness is a political project that continually changes the geographies of conservation, where virgin forests represent a proxy of this transformation (Iordachescu, 2021).

Beyond constituting a hot public debate for several years, 'virgin forests' have been the object of detailed political discussions ranging from national security to the development of big infrastructural projects (Wild Europe, 2018). It is important to contextualise this process within a broader eastern European interest for the protection of old-growth forests as part of a sustained international effort to identify and find ways of conserving wild nature under strict protection mechanisms. The Romanian legal framework for protecting virgin forests started to be developed only after the party states of the Carpathian Convention signed the Protocol on Sustainable Forest Management in May 2011. The Forest Protocol follows up on Article 7 (paragraph 5) of the Convention and refers to the designation of virgin forests and the need to protect them strictly. During the technocratic rule, ministerial ordinances set the criteria and detailed the instruments suitable for the strict protection of these iconic wild values (Ministry of Environment, 2017).

Along with strong governmental support for the definition, mapping and strict protection of old-growth forests as ecosystems "developing without any direct or indirect human influences" (Ministry of Environment, 2012), the technocratic period was marked by an explicit endorsement of a private initiative aiming to turn large areas of the Southern Carpathians into a wilderness reserve. Popularised in the media and political discourse as the 'European Yellowstone', this initiative is emblematic for the current transformations of nature protection in eastern Europe. Enclosing nature for the protection of biodiversity, whether by public or private actors, has international ramifications and is considered by many to be a global phenomenon (Peluso and Lund, 2011; White et al., 2012; Corson and MacDonald, 2012). The phenomenon is considered a sort of green grabbing, and it supposedly takes nature out of an extractivist logic and reserves it for ecotourism and the development of green businesses that are purported to be friendlier to the environment (Fairhead et al., 2012; Ojeda, 2012).

The Carpathia Project is a representative example for understanding how wild nature emerges within the region as a cheap resource. In the aftermath of the Romanian forest restitution, the proponents of the country's most iconic wilderness conservation project wrote to an international audience that

> Private owners want to sell, and what happens after a sales contract is rather irrelevant to these new owners. What if conservation organisations step dynamically into the picture? (Promberger and Promberger, 2015: 245)

Similarly, proponents of rewilding approaches advocate explicitly for the artificial cheapening of land to promote conservation initiatives:

> We propose to disconnect subsidies for marginal land from farming activities. Doing so will make farming less economical to owners of marginal land, which will reduce land prices, and hence reduce competition for land with other societal players, bringing opportunities for ecosystem restoration (Merckx and Perreira, 2015: 99)

A World-Class Wilderness Reserve

The Romanian case shows that after the cheapening of nature and its subsequent securitisation, ecotourism is frequently advanced as the silver bullet for many types of problems, from habitat degradation to land abandonment and rural poverty. So far, ecotourism has been presented to the general public as the only development alternative possible, as it is a fair economic model not only for nature but also for locals. In the Făgăraș Mountains, part of the Southern Carpathians, ecotourism initiatives have been sustained by a logic of securitisation. This captures the processes of capital accumulation as they are bound to a vast array of resource enclosures and dispossessions (Kelly and Ybarra, 2016; Masse, 2016; Masse and Lundstrum, 2016; Huff and Brock, 2017). The Carpathia Project is a good illustration of various processes at play in the creation of the eastern European wilderness frontier: the project is proposed by a conservation foundation that secures an entire territory for future accumulation by concomitantly taking over the roles of exclusive owner, custodian of Natura 2000 sites, administrator of hunting grounds, and as a member of historical forest commons. The project aims to be an example and blueprint for future initiatives in the region (Promberger, 2019). This case is relevant not only for its pioneering vision, but also for its ambition to become a model for other initiatives on the continent. FCC's founders are the leaders of the wilderness movement in Europe. Some of them pursue their rewilding projects; others put great efforts into lobbying for wilderness at the EU institutional level.

In this light, the wilderness conservation project acts as a veritable new frontier of land control. While enclosures have a long history in Europe and elsewhere, this specific enterprise stands out through its mechanisms. Peluso and Lund (2011) appreciate that what is different in the contemporary wave of enclosures are the alliances backing the project, as well as its general economic rationale. The FCC's conservancy allegedly takes nature out of an extractivist commercial logic and includes it in a non-extractive circuit (for ecotourism or contemplation). In other words, the Carpathia Project is justified as an attempt to repair the harm done by humans (i.e. former owners) by giving the land back to nature (i.e. wilderness). This new way of drawing boundaries between the human and the wild is seen here as a 'territorialisation' process (Peluso and Lund, 2011: 668). As this process includes dispossession, rights transfer and securitisation of resources, its losers end up being pushed towards a 'systemic edge', where expulsion from the economic, social and natural landscape is so advanced that it becomes hardly visible (Sassen, 2015).

Foundation Conservation Carpathia (FCC) is the most important private conservation actor in Romania and aims to be a leading example at the European level. Over the last ten years, the FCC has sought to protect and restore large forested areas in the eastern part of the Southern Carpathian Mountains. Their approach has centred on using private and public money to buy as much land as possible and ensure its full protection. Leasing hunting rights, acquiring custody of Natura 2000 protected areas and cataloguing virgin forests complemented their approach towards the strict protection of an allegedly untouched nature. By the end of 2019 the foundation and its commercial companies owned and administered over 22,000 hectares of forests and alpine pastures, and are considered one of the biggest private forest owners in the country. Besides buying land for strict protection, the foundation has acquired the custody of two Natura 2000 protected areas, a further almost 15,000 ha. Another strategy to ensure strict protection of wildlife within the project area was to bid for and acquire exclusive hunting rights. Buying land, being a custodian of protected areas and managing the hunting grounds are the strategies through which the FCC builds the future 'European Yellowstone'. If Africa has Serengeti and Kruger, and North America has Yellowstone, the time has come for Europeans to have their own, emblematic Yellowstone. This comparison is not fortuitous, however 'the European Yellowstone' has become common parlance among conservationists and nature lovers alike, frequently being adopted by policy-makers (Ziare.com, 2016; Rear, 2018).

Carpathia is a concrete example of how and where internationally discussed ideas about the strict protection of wild nature, understood as separated from human use, are put into practice. To achieve conservation objectives, the FCC proposed and followed two strategies. First, it worked to restore forest and aquatic ecosystems by reforesting barren slopes, covering old eroded forestry roads and reconstructing riparian alder habitats (*Alnus incana*). A generous LIFE+ grant and several other projects contributed to the successful implementation of this approach, resulting in more than 1.8 million trees being planted. Second, the foundation aims to reintroduce two missing species, considered of great value for an area aspiring to be a 'world-class reserve'. The beaver (*Castor fiber*) and the European bison (*Bison bonasus*) are the usual suspects in many rewilding projects on the continent, and scientists have devoted particular attention to the practicalities of these projects (Tănăsescu, 2017; Tănăsescu, 2019; Vasile, 2018b). Here they are expected to recreate mosaic landscapes and restore the natural ecosystems as they are abandoned or are not adequately managed. A grant of 5 million USD, awarded to the FCC by the Cambridge Conservation Initiative in early March 2019, is currently dedicated to this process (Endangered Landscapes Programme, 2018).

If in the early years Carpathia was merely a project aimed at stopping illegal logging around Piatra Craiului Massif, it has evolved over the years, at times with the state's help, into an enterprise for creating an iconic national park around the Făgăraș Mountains, considered the last unfragmented mountain range in Europe. The areas that the foundation currently controls are expected to constitute the strictly protected core of the future national park. At the peak of its governmental support the park was expected to be operational by the end of 2020.

Such a daring project would not be possible without direct governmental endorsement manifested in moral and legal support (Iordachescu, 2018). Over the last decade, important political figures from various parties have shown their appreciation for the wilderness reserve. This peaked in 2016 when Romania was ruled for about one year by a cabinet of technocrats led by the former EU Commissioner for Agriculture and Rural Development, Dacian Cioloș. Previously, in February 2014, the FCC received for the first time confirmation that the central authorities backed their project. The liberal Lucia Varga, holding office at the Ministry of Environment and Climate Change, signed a collaboration protocol with the foundation offering them full technical support for stopping the illegal logging and developing the conservation initiative in Făgăraș.

Towards the end of 2015, an unpredictable change of executive power took place in the country. The social democrats, led by Victor Ponta, resigned in the middle of a massive corruption scandal that triggered large public demonstrations. President Klaus Iohannis invited the non-affiliated Dacian Cioloș to form a government until the next parliamentary elections. Two FCC board members were appointed as state secretaries in the Ministry of Environment. Upon taking office, both of them announced an interruption of their roles in the FCC for the period of their appointment.

Two months after her appointment, the Minister of Waters and Forests went on an official visit over Făgăraș accompanied by the FCC's directors. The trip, financed by the FCC, also included the BBC's *Wild*

Carpathia documentary presenter Charles Ottley, and involved several other national celebrities in ongoing environmental campaigns. Both the minister and the FCC posted social media pictures of deforested mountain slopes in the middle of endless virgin forests. They reminded their followers about the urgency to save these wonders by supporting the creation of the 'European Yellowstone'.

Later that year, in September 2016, the government announced publicly that a new memorandum for establishing the Făgăraș Mountains National Park had been proposed for public consultation. The very first page of this official document advertised the proposed park as Europe's own Yellowstone: "Thus, Făgăraș Mountains National Park could become the most important national park of Europe regarding its rich biodiversity and extended area, a veritable 'European Yellowstone'" (Guvernul României, 2016: 1). The public consultation, on the other hand, did not go as expected, so for the time being the 2020 target for the park being operational remains a missed target.

According to the document, the proposed development vision for the area revolves around green businesses and ecotourism enterprises such as low impact visitations, wildlife watching facilities and animal tracking tours catering to an affluent Western audience. The locals are expected to propose business plans and develop their initiatives under the direction of the FCC and its partners, Conservation Capital, Romanian Association for Ecotourism, and others (Iordachescu and Vasile, 2016). Extractive processes such as commercial logging, domestic grazing, foraging and other traditional land uses are mainly excluded from this vision.

Securing and controlling access to local resources, enclosing the commons and commodifying charismatic wildlife are facets of this attempt that draw strict boundaries between 'wild' and 'domesticated' nature around the Făgăraș Mountains. Not everyone has experienced the same impact on their livelihoods by the strict conservation regime. Most of the villagers who privately own pastures and forests felt the arrival of the FCC to a lesser extent. At the same time, Roma communities, who possess no land, have no stable jobs, and live in precarious settlements, felt the impact the most. Between these two polarised categories are the shepherds, farmers, foresters, guesthouse owners, hunters, and many others who either had asked for their interests to be represented by the local authorities or opposed the foundation directly themselves.

Dispossession and Exclusion in the 'European Yellowstone'

Yellowstone is an important brand within the global conservation movement: while the park played an essential role in framing the spectacle of wildlife as part of a standardised, commodified experience (Rutherford, 2011), its foundation was marked by brutal dispossessions and genocide (Cronon, 1996).

The most decisive impact of the Carpathia Project, the 'European Yellowstone', has been felt in the historical region of Muscel, situated on the southern side of the Făgăraș Mountains. Most of the municipalities here are composed of several villages whose agricultural lands and forests extend from the hills to the alpine pastures. Historically, the area was relatively well off, situated between the first two capitals of the medieval Principality of Wallachia. Animal husbandry, forest exploitation, and commerce across the mountains between Transylvania and Wallachia have been the basis of this region's economic development. As almost all villages retained their privileges from medieval to modern times, the landscape and most natural resources have been governed by commons and customary rule until the land was nationalised from 1948 onwards (Vasile, 2018a). From 2000, a new restitution law allowed former historical owners to take back their lands, so the common ownership of forests became a source of pride and collective action throughout the region (Vasile, 2009). Thus, locals' strong opposition towards the wilderness conservation project did not come as a surprise for anyone, and the first years of the project were marked by rumours and suspicion rather than by open consultations and dialogue.

Aside from various forms of everyday resistance that never morphed into organised violent revolt, there have been different types of mobilisation by local authorities concerning rumours about declaring Făgăraș Mountains a national park. Although the state government's memorandum mentioned that the population around the Făgăraș Mountains was 73,000 inhabitants, it did not organise any consultation meeting before or after the memorandum was made public. The document's preamble read:

> The national park could attract over 500 million potential visitors from Europe. [...] through the establishment of Făgăraș Mountains National

> Park, the local communities surrounding the mountains have the unique opportunity of making it to the international map of tourism (Guvernul României, 2016, translation my own)

Rather than flattering local authorities, these words infuriated them. In November 2016 a big meeting was organised in Șercăița, a village on the northern side of Făgăraș. Representatives of thirty-three commons were joined by twelve mayors who discussed the memorandum and reaffirmed their opposition to the FCC's plans to build a 'world-class wilderness reserve'. Together they signed the Resolution of Șercăița, an official document that was submitted to the technocratic government. In four points, they asked the government to stop the establishment of the national park and to respect their property rights as granted by the Romanian constitution. They also filed a complaint to the National Anti-Corruption Office in which they accused the government of adopting a private conservation project as a state project of public interest.

Fig. 2. Rudari permanent settlement. Photograph by George Iordăchescu (2018).

Another important group that has never been at the negotiation table despite being directly impacted by the development of the wilderness reserve is that of local Roma communities. In many hilly or mountainous regions of Romania, different groups of Roma (calling themselves *Rudari*) were engaged in patron-client relations around forest exploitation, precarious agricultural work, scrap iron collection and other types of informal livelihoods that proliferated during the post-socialist period

(Dorondel, 2009; Dorondel, 2016). Around the Făgăraș Mountains, these realities were not different. All seven Rudari communities that I visited were economically deprived compared to nearby villages, in terms of infrastructure and public amenities. An unclear land tenure situation has been doubled here by precarious living conditions sometimes involving a lack of safe drinking water, or a heightened probability of flooding with the advent of severe rains.

For many of the interviewed families, their livelihoods were seriously affected after their access to areas rich in mushrooms or to nearby forests was halted. Until recently, they enjoyed customary access to these resources. It is here where everyday forms of resistance were most frequently performed: Rudari's weapons of the weak involved an entire set of actions, from petty firewood stealing to regularly breaking the barriers and fences installed by the FCC. They have often been fined, their carts, horses and chainsaws seized, and they were sometimes beaten, or even imprisoned, by gendarmes. Most of the clashes were with the rangers employed by the foundation to patrol the valleys alone or accompanied by gendarmes. These clashes happened inside and outside the protected forests.

During the last two years, the FCC has radically changed their public relation strategy towards greater openness and inclusivity. They have been very active in promoting their plans at local folklore festivals and even organised a Forest Carnival for 300 guests in Rucăr in 2018. Regardless of these attempts, locals' mobilisation against the project has remained strong. As the FCC started a set of consultative meetings in April and May 2019, people gathered in significant numbers in Râmnicu Vâlcea, Sibiu and Brașov to express their concerns. Farmers and the presidents of commons particularly voiced profound disagreement with FCC plans. On the western side of the Făgăraș Mountains, as well as on the eastern side, people have a recent history of conflicts with the administrations of Cozia and Piatra Craiului National Parks, both established in the mid-2000s without adequate public consultation. The discussions during the meetings convened by the foundation in May 2019 revolved more around the fears about future restrictions than around issues related to the value of wildlife and the ecosystem services offered by the future national park. As they have been reported by the local media, none of the meetings ended in a constructive way (Nostra

Silva, 2019a; Nostra Silva, 2019b). As the summer started, Barbara Promberger, executive director of the FCC, was invited to the National Geographic Explorers Festival in Washington. Here she spoke about how the foundation puts great effort into improving local communities' economic situation but finds nothing but suspicion and distrust.

Being the ones to bear the costs of wilderness preservation, locals fear that timber will be scarcer, grazing areas less bountiful and that conflicts with wild animals will increase. As they are offered promises of significant gains from the development of ecotourism, they also have their own ideas and concepts about how tourism should be developed in the area. Many locals, both persons with decision-making power as well as guesthouse owners and small farmers, believe that mass tourism and resorts with winter sports facilities would be more beneficial for the economies of their villages.

All of these forms of contention should not be seen as a rejection of nature protection or as a disinterest or aversion to environmental issues. People in the area feel a deep attachment to their mountains. Through the historical institutions of commons, natural resources have been used and managed in a sustainable way for centuries. These concerns should be interpreted as a disapproval of a top-down, strict conservation approach that attempts to 'save' a nature that is unknown and unrecognisable to those who live there—the wilderness and its narratives are totally separate from traditional use and local history.

Conclusion

As intact landscapes, old-growth forests and strictly protected wildlands are considered an essential element in recent EU climate mitigation and biodiversity strategies (such as the New Green Deal), I see the development of wilderness protection in the region as a process of unequal ecological exchange between a wealthier, Western core and a periphery, where the decision-making processes, hegemonic conservation knowledge and financial mechanisms of the former are concentrated and deployed to fix, restore, reconstruct and sustainably use the 'nature' discovered in the latter, which is characterised by backwardness, subsistence, land abandonment and depopulation. This process unfolds as the creation of a green internal periphery, mainly to achieve EU and member states' ambitions to sustain green economic growth and lead the global fight against climate change.

These various wilderness conservation projects are bound together not only by strong political and ideological support, but also by the similar socio-economic local contexts that enable them. Local conditions such as declining rural population, actual or relative land abandonment, the demise of traditional land-use practices and 'cheap' nature, are all features of this new green internal periphery represented by the eastern European wilderness frontier.

References

Adams, W., et al., (2014). New spaces for nature: re-territorialisation of biodiversity conservation under neoliberalism in the UK. *Transactions of the Institute of the British Geographers*, 39(4), 574–88. https://doi.org/10.1111/tran.12050.

Bastmeijer, K., 2016. Introduction, in: Bastmeijer, K. (Ed.), *Wilderness Protection in Europe*. Cambridge University Press, Cambridge, pp. 3–37. https://doi.org/10.1017/cbo9781107415287.001.

Bauer, N., et al., 2017. Attitudes towards nature, wilderness and protected areas: a way to sustainable stewardship in the South-Western Carpathians. *Journal of Environmental Planning and Management*, 61(5–6), 857–77. https://doi.org/10.1080/09640568.2017.1382337.

Blavascunas, E., and Konczal, A., 2019. Bark-beetles and ultra-right nationalist outbreaks, in: Brain, S. and Pal, V. (Eds.), *Environmentalism under Authoritarian Regimes: Myth, Propaganda, Reality*. Routledge, New York, pp. 96–122. https://doi.org/10.4324/9781351007061-6.

Corson, C., and MacDonald, K., 2012. Enclosing the global commons: the convention on biological diversity and green grabbing. *The Journal of Peasant Studies*, 39(2), 263–83. https://doi.org/10.1080/03066150.2012.664138.

Cronon, W., 1996. The Trouble with Wilderness: Or, Getting Back to the Wrong Nature. *Environmental History* 1(1), 7–28. https://doi.org/10.2307/3985059.

Dorondel, S., 2016. *Disrupted Landscapes. State, Peasants and the Politics of Land in Postsocialist Romania*, Berghahn Books, New York. https://doi.org/10.2307/j.ctvgs0brw.

——, 2009. They should be killed. Forest restitution, ethnic groups and patronage in post-socialist Romania, in: Fay, D., and James, D. (Eds.), *The Rights and Wrongs of Forest Restitution: "Restoring What was Ours."* Routledge, London, pp. 43–65. https://doi.org/10.4324/9780203895498-11.

Egerer, H., 2016. Wilderness Protection under the Carpathian Convention, in: Bastmeijer, K. (Ed.), *Wilderness Protection in Europe*. Cambridge University Press, Cambridge. https://doi.org/10.1017/cbo9781107415287.010.

Endangered Landscapes Programme, 2018. 'Eight landscapes for life announced', 8 October, https://www.endangeredlandscapes.org/2018/10/08/eight-landscapes-for-life-announced/.

——, 2017. 'Vision', 29 September, https://www.endangeredlandscapes.org/wp-content/uploads/2017/09/Vision-document-29-Sept.pdf.

EuroNatur, 2016. 'Blue Heart in the European Parliament', 2 July, https://www.euronatur.org/en/what-we-do/campaigns-and-initiatives/save-the-blue-heart-of-europe/save-the-blue-heart-of-europe-news/detail/news/blue-heart-in-the-european-parliament/.

European Commission, 2020. 'EU Biodiversity Strategy for 2030', *COM(2020) 380 final*, 20 May, https://ec.europa.eu/info/sites/default/files/communication-annex-eu-biodiversity-strategy-2030_en.pdf.

——, 2013. 'Guidelines on wilderness in Natura 2000', Technical Report — 2013-069, https://ec.europa.eu/environment/nature/natura2000/wilderness/pdf/WildernessGuidelines.pdf.

European Green Belt, 2018. 'European Green Belt Initiative', https://www.europeangreenbelt.org/.

European Parliament, 2018. 'Large carnivore management plans of protection: best practices in EU Member States', Report commissioned by PETI Committee, 15 February, https://op.europa.eu/en/publication-detail/-/publication/29b11762-437b-11e8-a9f4-01aa75ed71a1.

——, 2009. 'Resolution on Wilderness in Europe', *P6_TA(2009)0034*, 3 February, https://www.europarl.europa.eu/sides/getDoc.do?pubRef=-//EP//TEXT+TA+P6-TA-2009-0034+0+DOC+XML+V0//EN.

Fairhead, J., Leach, M., Scoones, J., 2012. Green Grabbing: a new appropriation of nature? *The Journal of Peasant Studies*, 39(2), 273–61. https://doi.org/10.1080/03066150.2012.671770.

Fox, K., 2011. *Peasants into European Farmers? EU Integration in the Carpathian Mountains of Romania*. LIT Verlag, Zürich. https://doi.org/10.3726/978-3-653-01758-8/1.

Gross, M., 2016. Europe's Last Wilderness Threatened. *Current Biology*, 26(14), 641–43. https://doi.org/10.1016/j.cub.2016.07.009.

Guvernul, României, 2016. 'Guvernul a aprobat demersurile pentru declararea Muntilor Fagaras ca Parc National', *gov.ro*, 14 September, https://www.gov.ro/ro/guvernul/sedinte-guvern/guvernul-a-aprobat-demersurile-pentru-declararea-muntilor-fagara-ca-parc-national#null.

Gzeszczak, R., and Karolewski, I., 2017. Bialowieza Forest, the Spruce Bark Beetle and the EU Law Controversy in Poland, *Versassungsblog*, 27 November, https://verfassungsblog.de/bialowieza-forest-the-spruce-bark-beetle-and-the-eu-law-controversy-in-poland/.

Hochtl, F., et al., 2005. 'Wilderness': What it means when it becomes reality — A case study from the southwestern Alps. *Landscape and Urban Planning*, 70(1), 85–95. https://doi.org/10.1016/j.landurbplan.2003.10.006.

Huff, A., and Brock, A., 2017. Accumulation by Restoration: Degradation Neutrality and the Faustian Bargain of Conservation Finance. *Antipode Online,* 6 November, https://antipodeonline.org/2017/11/06/accumulation-by-restoration/.

Iordachescu, G., 2021. Becoming a virgin forest: From Remote Sensing to Erasing Environmental History. *Environment & Society Portal Arcadia,* 2021(10). http://www.environmentandsociety.org/arcadia/becoming-virgin-forest-remote-sensing-erasing-environmental-history.

——, 2020. Romania's forests and a global pandemic: a case of ecological entanglement. *Biosec Project*, 5 May, https://biosecproject.org/2020/05/05/blog-romanias-forests-and-a-global-pandemic-a-case-of-ecological-entanglement/.

——, 2018. Making the European Yellowstone — Unintended Consequences or Unrealistic Intentions? *Environment & Society Portal Arcadia*, 2018(10). https://doi.org/10.5282/rcc/8303.

——, 2014. 'Inventing Traditions for Whom? Commodification of Authenticity and Failed Rural Development in Capitalist Romania'. Central European University, Budapest (Unpublished MA thesis).

Iordachescu, G., and Vasile, M., 2016. 'The socio-economic context of the communities neighbouring the Fagaras mountains', Report submitted to Foundation Conservation Carpathia, https://www.carpathia.org/wp-content/uploads/2019/11/ECOSS-Raport-A2-Studiu-socio-economic.pdf.

Jørgensen, D., 2015. Rethinking Rewilding. *Geoforum*, 65, 482–88. https://doi.org/10.1016/j.geoforum.2014.11.016.

Kelly, A., and Ybarra, M., 2016. Introduction to themed issue: Green Security in Protected Areas. *Geofurm*, 69, 171–75. https://doi.org/10.1016/j.geoforum.2015.09.013.

Kirchoff, T., and Vicenzotti, V., 2014. A Historical and Systematic Survey of European Perceptions of Wilderness. *Environmental Values*, 23(4), 443–64. https://doi.org/10.3197/096327114x13947900181590.

Kupper, P., 2014. *Creating Wilderness: A Transnational History of the Swiss National Park*. Berghahn Books, New York.

Lawrence, A., and Szabo, A., 2005. Forest restitution in Romania: Challenging the value system of foresters and farmers. *Human Ecology Working Paper*, 5 Jan, Environmental Change Institute, University of Oxford.

Lupp, G., et al., 2012. Wilderness. Consequences of a mental construct for landscapes, biodiversity and wilderness management. *European Journal of Environmental Sciences*, 2(2), 110–14. https://doi.org/10.14712/23361964.2015.31.

——, 2011. Wilderness. A Designation for Central European Landscapes. *Land Use Policy*, 28, 594–603. https://doi.org/10.1016/j.landusepol.2010.11.008.

Martin, V., et al., 2008. Wilderness Momentum in Europe. *International Journal of Wilderness* 14(2), 34–43.

Masse, F., and Elizabeth, L., 2016. Accumulation for securitization: Commercial poaching, neoliberal conservation and the creation of new wildlife frontiers. *Geoforum*, 69, 227–37. https://doi.org/10.1016/j.geoforum.2015.03.005.

Masse, F., 2016. The Political Ecology of Human-Wildlife Conflict. *Conservation and Society*, 14(2), 100–11. https://doi.org/10.4103/0972-4923.186331.

Merckx, T., and Pereira, H., 2015. Reshaping agri-environmental subsidies. From marginal farming to large-scale rewilding. *Basic Applied Ecology*, 16, 95–103 https://doi.org/10.1016/j.baae.2014.12.003.

Ministry of Environment, 2017. 'Order 2525 of 30/12/2016', *Monitorul Oficial*, 63 of 25 January 2017.

——, 2012. 'Order 3397 of 10/09/2012', *Monitorul Oficial*, 668 of 24 September.

Moore, J. W., 2015. *Capitalism in the Web of Life: Ecology and the Accumulation of Capital*. Verso, London.

Moore, J., and Patel, R., 2018. *A History of the World in Seven Cheap Things*. Verso Books, London.

Nostra Silva, 2019a. Familia Promberger si Conservation Carpathia fata in fata cu mosnenii devalmasi. *Nostra Silva*, 16 May, https://www.nostrasilva.ro/media/familia-promberger-si-conservation-carpathia-fata-fata-cu-mosnenii-devalmasi/.

——, 2019b. Conservation Carpathia: politica bocancului pe grumazul proprietarului si a pumnului in gura presei. *Nostra Silva*, 24 May, https://www.nostrasilva.ro/activitati/conservation-carpathia-politica-bocancului-pe-grumazul-proprietarului-si-pumnului-gura-presei/.

Ojeda, D., 2012. Green Pretexts: Ecotourism, Neoliberal Conservation and Land Grabbing in Tayrona National Park in Colombia. *The Journal of Peasant Studies*, 39(2), 357–75. https://doi.org/10.1080/03066150.2012.658777.

Peluso, N. L., and Lund, C., 2011. New frontiers and land control: Introduction. *The Journal of Peasant Studies*, 38(4), 811–36. https://doi.org/10.1080/03066150.2011.607692.

Petrova, S., 2014. *Communities in Transition: Protected Nature and Local People in Eastern and Central Europe*. Routledge, London. https://doi.org/10.4324/9781315572949.

Promberger, C., 2019. Carpathia — A world class vision proposed for the Carpathians, *Wild Europe Conference*, 20–21 November, Bratislava.

Promberger, B., and Promberger, C., 2015. Rewilding the Carpathians. A present-day opportunity, in: Wurther, G., et al. (Eds.), *Protecting the Wild: Parks and Wilderness, the Foundation for Conservation*. Springer, Dordrecht. https://doi.org/10.5822/978-1-61091-551-9_25.

Rear, J., 2018. The Yellowstone of Europe: Plans for a major national park in Romania, *Verdict*, 15 May, https://www.verdict.co.uk/fagaras-national-park-romania/.

Romanian Presidency of the EU Council, 2018. Photo exhibition: Wild Heart of Europe. *romania2019.eu*, 7 January, https://www.romania2019.eu/event/photo-exhibition-wild-heart-of-europe/.

Rewilding Europe, 2020. 'European Rewilding Network', https://rewildingeurope.com/european-rewilding-network/.

Rutherford, S., 2011. *Governing the Wild: Ecotours of Power*. University of Minnesota Press, Minneapolis. https://doi.org/10.5749/minnesota/9780816674404.001.0001.

Saarinen, J., 2015. Wilderness use, conservation and tourism: what do we protect and for and from whom? *Tourism Geographies*, 18(1), 1–8. https://doi.org/10.1080/14616688.2015.1116599.

Sassen, S., 2015. At the Systemic Edge. *Cultural Dynamics*, 27(7), 173–81. https://doi.org/10.1177/0921374014567395.

Schumacher, H., 2008. More Wilderness for Germany: Implementing an important objective of Germany's National Strategy for Biodiversity. *Journal for Nature Conservation*, 42, 45–52. https://doi.org/10.1016/j.jnc.2018.01.002.

Schwartz, K., 2006. *Nature and National Identity after Communism. Globalizing the Ethnoscape*. University of Pittsburgh Press, Pittsburgh.

Tănăsescu, M., 2019. Restorative Ecological Practice: The case of the European Bison in the Southern Carpathians, Romania. *Geoforum*, 105, 99–108. https://doi.org/10.1016/j.geoforum.2019.05.013.

——, 2017. Fieldnotes on the Meaning of Rewilding. *Ethics, Policy & Environment*, 20(3), 333–49. https://doi.org/10.1080/21550085.2017.1374053.

Schlingemann, L., et al., 2017. Combating Wildlife and Forest Crime in the Danube-Carpathian Region, *A UN Environment—Eurac Research—WWF Report*, https://wedocs.unep.org/bitstream/handle/20.500.11822/22225/Combating_WildlifeCrime_Danube.pdf?sequence=1&isAllowed=y.

UNESCO, WHC. Decision 41 COM 8B.7, https://whc.unesco.org/en/decisions/6879.

Urban, S., 2016. Wilderness protection in Poland, in: Bastmeijer, K. (Ed.), *Wilderness Protection in Europe*. Cambridge University Press, Cambridge. https://doi.org/10.1017/cbo9781107415287.018.

Vasile, M., 2020. The Rise and Fall of a Timber Baron: Political Forests and Unruly Coalitions in the Carpathian Mountains of Romania. *Annals of the American Association of Geographers*, 110(6), 1952–68. https://doi.org/10.1080/24694452.2020.1723399.

——, 2019. Forest and pasture commons in Romania. Country Report, ICCA Consortium, https://www.iccaconsortium.org/index.php/2019/07/22/forest-and-pasture-commons-in-romania/.

——, 2018a. Formalizing the commons, registering rights. The making of the forest and pasture commons in the Romanian Carpathians. *International Journal of the Commons*, 12(1), 170–201. https://doi.org/10.18352/ijc.805.

——, 2018b. The Vulnerable Bison: Practices and Meaning of Rewilding in the Romanian Carpathians. *Conservation and Society*, 16(3), 217–31. https://doi.org/10.4103/cs.cs_17_113.

——, 2009. Corruption in Romanian Forestry. Morality and local practice in the context of privatization. *Revista Romana de Sociologie*, 20(1–2), 105–20.

White, B., et al., 2012. The New Enclosures: Critical Perspectives on Corporate Land Deals. *The Journal of Peasant Studies*, 39(3–4), 619–47. https://doi.org/10.4324/9781315871806.

Wild Europe, 2019. New initiatives from a decade of progress. A conference for an action plan to protect and restore wilderness and old growth/primary forest in Europe. 20–21 November, Bratislava.

——, 2018. Old Growth Forest Protection Strategy Outline, https://www.wildeurope.org/wp-content/uploads/2019/10/old-growth-forest-protection-strategy-outline.pdf.

WILD10, 2015. *Vision for a Wilder Europe.* 2nd edn, http://wild10.org/wp-content/uploads/2015/03/WILD10-Vison-for-a-Wilder-Europe-March-2015-Final-print.pdf.

WWF Romania, 2018. 2000 de hectare de paduri virgine sunt securizate prin eforturile WWF-Romania. *wwf.ro*, 9 February, https://www.wwf.ro/resurse/comunicate_de_presa/?uNewsID=323031.

Ziare.com, 2016. Muntii Fagaras ar putea devein cel mai mare parc natural din Europa. *Ziare.com*, 24 February, https://m.ziare.com/stiri/muntii-fagaras-ar-putea-deveni-cel-mai-mare-parc-natural-din-europa-1410212.

9. Domesticating the Taste of Place
Post-Socialist *Terroir* and Policy Landscapes in Tokaj, Hungary

June Brawner

[...] wine tasters must draw objective conclusions about a wine from their subjective responses to it, and wine-makers must create conditions they hope will produce a certain taste for us.

Smith, 2013: 114

Terroir as Political Ecology

"For the People to Drink": Bittersweet Change

As part of extended fieldwork in Hungary, I attended a formal wine tasting on the theme of indigenous *furmint* grape wines in the summer of 2017. At a cosy, English-language culinary centre, British wine writer Joseph[1] is leading the group (a mixture of about eighteen Hungarian locals and North American visitors) into the *aszú* portion of the wine list. *Aszú* wines are a traditional specialty of the Tokaj wine region, made from grapes that have been affected with the fungus *Botrytis cinerea*. The fungus, which arrives in late autumn (if at all), settles on the berries, then vampirically consumes the flesh from the outside, changing the composition of the juice as it reinjects the fruit with its waste products. The result of this "desirable meeting between a fruit and a fungus" (Magyar and Soos, 2016: 31) is irreproducible: a highly sugared,

1 Names have been changed.

 https://doi.org/10.11647/OBP.0244.09

complexly flavoured, and deeply *concentrated* taste that showcases the wine's unique growing conditions.

Joseph begins the *aszú* tasting with a caveat I had heard in similar settings: "You know, I'm not a sweet wine fan at all, but I—I don't really consider Tokaji *aszú* a sweet wine, I just consider it a very *rich* wine". He says the high acidity of the indigenous furmint grape in the wine balances the sweetness resulting from the botrytis affectation. Where it would otherwise taste "cloying", it is—Joseph explains—instead something *else*. He asks a leading question, "Does it taste that sweet, or is it *complex*?"

The silence of the room is interrupted by a Hungarian woman: "I have an interesting question. Why is it that most people don't like sweet wines? Why is it that—that you just *also* mentioned it?" Joseph raises his eyebrows and becomes jokingly defensive, "Ah, but I would drink *that*!" he insists, gesturing to the bottle of *aszú*. "I don't know," he replies, "because fashions change, I think. Earlier, sweet wines were, like, massively popular." "Right!" The Hungarian woman urges him on. He continues, "Well, certainly, ehh—one or two generations ago it was seen as a kind of—you know—*luxurious* thing to sit there and absorb lots of sugar. I think now that people are like, you mention sugar and people freak out." "But don't people drink wine with soda water," she asks, "because it can take the sweetness out of it?" She is referring to *fröccs*, a traditional spritzer made with soda water and wine.

Joseph acknowledges the popularity of the *fröccs*, and even that he likes them occasionally, and the woman returns to her original questioning: "But in Hungary, it was—people didn't drink that expensive wine. They didn't *make* expensive wine before. It was for the *people* to drink. When it's made at home...that's the history. It has changed now." She thinks aloud of a time when her family used to purchase and drink wine regularly, when it was less expensive, and less formally produced and consumed. Contrasting the historic production of homemade wine (for local trade and consumption) with the new wave of internationalisation and formal exchange of taste knowledge, she explains, "They make something *better* now. *We* used to buy and drink wine, when we could make wine at home."

Denaturalising the 'Taste of Place'

The Hungarian attendee at the furmint tasting is emblematic of a rift I witnessed during my fieldwork in Budapest and Tokaj, Hungary. Like other ethnographic accounts of post-socialist transitions, I saw that changes were not merely structural, but involved reorientation to a new world of capitalism—a reorientation that encompasses everyday practice and experiences, and a realignment and valuation of 'familiar' tastes. In products like wine, 'good tastes' are associated with 'good places': the best *terroirs*. Because tasting *terroir* requires translational narratives that link ecologies to taste qualities, this has put the 'unknown *terroir*' of Central and Eastern Europe (CEE) at a disadvantage against contemporary Western European and New World offerings (see Jung, 2014).

Until recently, taste in agri-food chains has been a "monolithic and largely externally defined evaluation of a product" (Demossier, 2018: 106) with little attention paid to the direct influence of aimed-for tastes on producer labour: how the logic of production systems, motivated by the tastes of the intended recipients, plays out on the material landscape. In Tokaj, however, it is not so far-fetched to imagine the direct link between the Russian 'palate' and mid-twentieth-century production practices. The conspicuous consumption of once-elite Tokaji wines, then aged for years in oak barrels, encouraged the Tokaji oak and coopering industry. It is no coincidence that the width of many (or most) Tokaji vineyard rows are the width of Russian tractors, having been replanted with French varietals in wide rows to allow for mechanisation and the mass market in the mid-twentieth century.

As Ulin (2013) convincingly explains, *terroir* must be denaturalised to be properly understood. In other words, while *terroir* has been employed to "unwittingly conceal and marginalize the historicity of social relations upon which the production and consumption of wine is based" (67), I use this chapter to make those connections explicit, connecting contemporary tensions in the wine world—often indexed through tastes and politics, as in the opening vignette—to the material environment: sites of ecologically embedded tastes (Krzywoszynska, 2015).

Originally catering to export markets and tastes abroad, twentieth-century communist production focused on a different audience—one internal to the Eastern Bloc. Today, the revival of Tokaj as an exclusive *terroir* renders its ecologies as monopolies of 'quality', albeit by evolving and contested definitions. Tokaji wines are naturalised as products of unique ecosystems, their unique tastes emphasised to make them paradoxically more like their Western, international counterparts. To focus on one aspect of this change, I emphasise the shift from local production and preferences for sweet, *aszú* wines to the contemporary trend for 'dry' wine. Because these two styles require commitments to very different growing and production methods, they represent a clear example of the relationship between ecologically embedded tastes and material political ecologies.

The result is a wine region of historic contradictions: where once there were sweet wines, international grape varietals, long ageing times in Hungarian oak barrels (for 'expensive' and aged tastes), bulk quantities, and sloped hills suited to mechanisation, today there is a push for dry/'mineral' wines, a select few indigenous grape varietals, brief ageing periods (for 'fresh', everyday tastes) in stainless steel tanks, limited quantities, and the return of terraced vineyards alongside hand-picking. As political ecologies, I will argue, vineyards are themselves relics of previous policy regimes and norms—the material consequences of capitalism (Peet and Watts, 1986), socialism, and a long history of producing the 'taste of place' where places reflect constantly changing political terrains.

Location, Location, Location

Hungary is a key wine-producing country in the CEE region, often dubbed the 'New Old World'. The first written record of wine production in Hungary dates to the fifth century CE. Perhaps because it is located on migration routes (situated between the origin of winemaking in the Southern Caucasus and continental Europe), Hungarian is one of only three languages in Europe in which the word for wine (*bor*) is not rooted in Latin (along with Greek and Turkish).

By the seventeenth century, winemakers in the Tokaj region of northeast Hungary determined that its best wines were derived somewhat consistently from a subset of special tracts; based on these patterns, they created the first vineyard classification system, put in writing by the 1730s. This involved dividing each vineyard tract (*dűlő*) into three quality classes based on several environmental and economic variables, helping to standardise the production of its primarily sweet wines. The region was then enclosed by royal decree in 1737, making Tokaj the second-oldest enclosed wine region in the world (Chianti, Italy predated this decree by forty-one years). Once dubbed the "Wine of Kings, King of Wines" by French royalty, Tokaj's international status all but vanished in the twentieth century following two world wars and four decades of communism (Liddell, 2003).

After a dynamic period of state-owned production during communism, Tokaji vineyards were systematically privatised through auction and voucher systems in the 1990s. Initially, old aristocratic estates were reformed (composed largely of first-class *dűlő*) and sold as units, many to foreign firms. The sale of agricultural land to foreigners was banned after 1994, and remains a point of contestation today (e.g. Brawner, 2021). Today, the official Tokaj region includes twenty-seven towns and villages in Borsod-Abaúj-Zemplén county and their surroundings (including a town also called Tokaj), as well as a smaller tract of disputed land in contemporary Slovakia (Figure 1). In contemporary Tokaj, a new generation of local and international

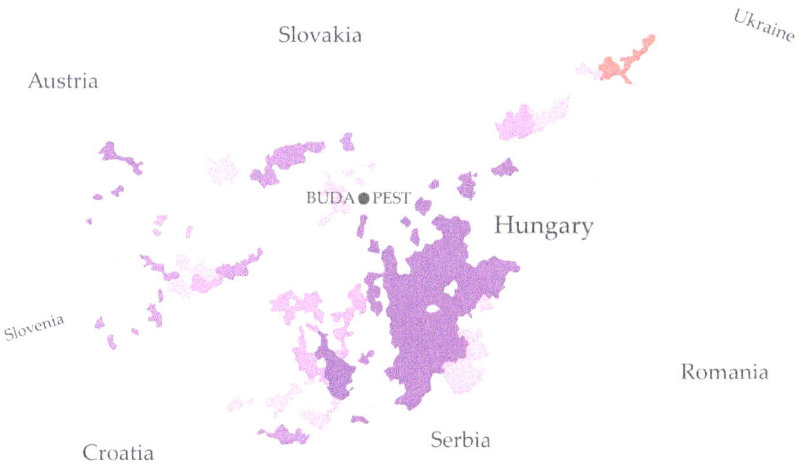

Fig. 1. Map of the wine regions of Hungary (purple), with the Tokaj wine region in red. Image by the author (2020).

winemakers seek to revive the once-popular "Wine of Kings" according to new ideals.

This chapter draws from ethnographic fieldwork conducted across several periods. Having lived in Hungary for several years prior, I returned for fieldwork and in the summers of 2014 and 2015, followed by an extended stay from 2016 to 2017. Returning in 2016, I used extensive participant observation, shadowing in wine-making and tasting spaces, archival research, environmental surveys, and formal interviews to explore the political ecology of *terroir*.

Regimes of Change

Legacies of Communist Production and Privatisation

Communist production left its mark in the viticultural systems of Tokaj. Writing in 2007, Hungarian viticulturalists Sidlovitz and Kator explain of the need for assistance from the EU:

> *The heritage of socialist viticulture is visible at the level of the vineyard management technique and the state of vineyard*, where one part is obsolete, the other is old, and the conversion proportion is [too weak for quality] wine production and quality improvement. (15, emphasis added).

The shift to 'quality' in winemaking implies not only higher prices fetched, but significantly higher costs of production due to the overwriting of communist 'landscapes' of production and the transformation of vineyards with wide, vertical rows into images of their historic predecessors. However, for Tokaji producers looking to maintain a living as winemakers, producing quantities of wine in the 'communist style' (which is, broadly speaking, created in accordance with local tastes) is a more guaranteed livelihood than reducing quantity and producing international styles that rarely reach the volumes or prices required for sustainable export. Beyond these markers, there is some evidence that regimes of over-fertilisation (intended to increase yield) have resulted in the presence of legacy nutrients in vineyard soils (Brawner, 2018).

The privatisation of Tokaj vineyards in the early 1990s was coupled with ecological consequences, such as "the fragmentation of vineyards designed and planted to be run as an integrated whole" (Liddell, 2003: 24). Former cooperative members might have received, for example, eight rows of vines on a twenty-hectare plantation; attempts to [re] consolidate production for private individuals were largely (and perhaps unsurprisingly) unsuccessful. Some others had inherited parcels in vineyards due to their lineage but had no interest or knowledge and so abandoned them. By 2001, wine writer Liddell observed that this led to the ad hoc use of pesticide sprays and partial vineyard abandonment with detrimental consequences: "when part of a vineyard is not properly looked after, the rest can only suffer" (2003: 24). "Just as collectivization solved the impracticality of running uneconomic units resulting from the breakup of estates in the late 1940s," he summarised, "the task now is to find a way of stitching broken-up vineyards back together again" (Liddell, 2003: 25). Integration contracts were introduced, renewable annually by the grower, to encourage producers to commit to harvest dates, sugar levels, volumes, etc., but with mixed results.

The serious costs associated with transforming a *dűlő* into operational vineyards of 'quality' *terroir* production has furthered the divide between locals with an interest in winemaking and entrepreneurs with capital, who are often foreign investors or members of the urban middle class. Many winemakers in the region are thus now located in Budapest or other centres, where they can afford a *dűlő* tract with their wages and a holiday home in one of the many near-empty village centres. For these producers, it is the love of the land and the hobby (often passed down from previous generations or inspired by mythologies of national heritage) that inspires their craft, which is rarely profitable in any conventional sense.

Impact of International Approaches

While multi-nationals bought large plantations in the early 1990s, some locals have more recently invested in the 'cast-offs' of these companies: adjacent areas that were too remote for large-scale production. Zoltán purchased one such tract in an original first-class *dűlő*:

> The first parcel we acquired was a small, 0.58-hectare area, offered to us by a larger company. They wanted to get rid of it as it was too far out for them [...] The area was in rundown condition with a traditional vine-stock cultivation; no one wants that these days. Winemakers generally dislike that, as it requires lots of manual labor.

> So the place was not appealing at all, riddled with fruit trees. However, when I saw the stone walls, saw this valley isolated from everything else, I knew inside that I must try and make this work.

He purchased these places with loans from friends and family, as well as some government aid. "Obviously, success wasn't handed to us. It wasn't like I just showed up, someone handed me four hectares of a *dűlő* and that was that. I had to establish relationships with local winemakers, slowly purchasing local areas". His scheme to turn the *dűlő* terrain from one of mass-production mode to quality, terraced areas meant that they have "used construction machines to clean out the terraces one by one". Because these areas are off-limits for tractors, they have also "managed to partner with an equestrian who is particularly handy with horses.... He also helps with horse-powered cultivator, plows, and hoes". Zoltán quantifies the cost of commitment to quality in his *dűlő*, some of which have only "2300 vines per hectare," while their "cultivation expenses are basically same as with 6250 vines per hectare". Given the recent political history in the area, it is unsurprising that the "most Hungarian" wine region struggles to unify its image.

Veteran winemaker Árpád was eager to explain every side of the political situation in Tokaj, which he insisted is inseparable from the history of the region. This includes, especially, the contention between insiders and outsiders: the style of winemaking (and thus rules and regulations associated with production) as debated amongst the newcomers and old-timers in Tokaj. While he does not entirely disapprove of foreign wineries in the region, he resents the concessions made by locals and transcribed into law on the basis of western European influence, explaining, "The French modified the Hungarian law for wine in 1994. They wrote it, actually, not the Hungarians. Hungarians typed it, but *they* thought it out".

Part of this new law, he explains, included the doing-away with wooden barrels. They "changed the old system, which entailed the following: as many *puttony* the wine had, plus two years extra for maturing. So, for example, a six *puttony aszú* was matured for eight years". His disdain for outsider influence underlines his retelling of these early years of privatisation, when "the French" arrived with new, modern methodologies to overwrite communist-era styles:

> Now the *aszú*'s maturing time is eighteen months [instead of two-to-eight years]. And I can't completely accept this, because I told them at so many meetings that we didn't become part of the [UNESCO] World Heritage because of the now-used, heated-cooled [stainless steel] tanks which generate 'uniwine' that they could make in Chile or France. What does this have to do with a Tokaji wine other than the raw material?
>
> The past 400 years were about the *aszú*... the wooden barrels, the cellars, the noble rot—this was the process. [...] And here they re-wrote our 400-year-old tradition. So, this is my opinion but also a fact. [...]
>
> The *aszú* wine has only its name now and the fact that you need *aszú* berries to make it.

For producers like Árpád, there is a sense that the revival of tradition in Tokaj was *more* possible during socialism: there may have been bulk, falsely aged wines shipped to the Russian market, but at least the "old ways" were maintained by locals who still aged *aszú* wines in barrels underground for years at a time. Today, liberalised markets require an even faster turnaround: a big task for the original *slow* wine. His anxiety about the future of Tokaj is parallel to his political concerns around Hungary and EU directives to accept minimum numbers of refugees during the recent refugee 'crisis': "Here, they say we should let in 100,000, because we have ten million people,"[2] he explains, "but I say that if this happens, Hungary disappears. Hungary is over. And they don't drink wine! Let's think this over: their kids won't, grandchildren won't, friends won't... You didn't think about this, did you?"

Today, Tokaji wine laws entail a common Protected Designation of Origin (PDO) label with standard production requirements. However, as producers like Árpád would argue, the heavy influence of foreign investors in the early 1990s led to entirely new rules—best practices that aim to create a product that is more suited to global taste trends: specifically, away from oxidation and "sweet" wines, and toward *terroir*-showcasing, dry wines (see also Brawner, 2019).

There are also broader political tensions at the European level. Hungary joined the EU in 2004, just as the EU was making strides toward draining the 'wine lake' that resulted from the consistent overproduction of wine in its member states in recent decades. In 2007, over 1.7 billion bottles were reported as surplus for several early-2000s

2 This refers to the quota system proposed by the European Parliament for refugee resettlement. I was unable to find any source that supported this 100,000 number.

vintages (Frank and Macle, 2007: 15), and "emergency distillation" (into industrial alcohol) became the fate of hundreds of millions of bottles of European wine each year (Wyatt, 2006). The post-productivist EU era and its rural development schemes are "defined by the buzzwords of multifunctionality, rural development, heritage and environmental concern" (Demossier, 2018: 136). This paradigm drives the promotion of rural landscapes as beds of artisan production and traditional methods in ways that simultaneously enhance localised foodways and rural tourism, while encouraging environmental conservation and biological diversity through specialty products (and, arguably, new forms of dependency on EU funds [e.g. Fischer and Hartel, 2012; Kovacs, 2019]).

Efforts to drain the 'wine lake' include EU vineyard 'grubbing-up' or vine pull schemes, initiated in the EU in 1988. Through these policies, producers with unprofitable vineyards may pull up their vines in exchange for cash payments. Thus, joining the EU in the early 2000s as a wine-producing country entailed much debate in and around Hungary regarding the requirements of new member states to comply with the strict production caps in place, entering into a single market already super-saturated with wine and with little interest in the contested wines of post-communist Hungary. If the communist era provided a steady market with little room for capitalising on quality, the new era has not offered the hoped-for replacement.

For all the political influence on Tokaji vineyards, the role of *taste* as a barrier to success emerges in nearly every discussion with foreign wine professionals, whether living in the region or appraising it as a visitor. One expert, from the US and now living in Tokaj, explained that it is local taste which holds the region back, as the old generation still insists on producing and drinking "gutter water". In another example, Liddell relates Hungary's sub-optimal wine production directly to local tastes as he notes (48):

> Finally—and sadly, because it continues to have a baleful influence on so much Hungarian winemaking—mention must be made of the Hungarian palate. Wine tastes are generally not at all sophisticated, and much wine is simply a vehicle for the alcohol it contains, as the small, dumpy glass usually used for drinking and tasting (filled to the brim) rather suggests. Your glass of wine is likely to be accompanied by a plate of *pogácsa* (small cheese scones).

Liddell's undeniably classist observation invokes Bourdieu's notion of taste (1984) as he observes that wine is merely a "vehicle" for local alcohol consumption. Indeed, the exclusivity of formal taste knowledge is obvious in these settings; perhaps a "good palate" is "proportional to the value and size of [their] wine cellar" (Teague, 2015). If *terroir* is undemocratically distributed geographically, so is the hierarchy of sensory knowledge required to appreciate it. While in previous eras, winemakers were considered the most knowledgeable judges of taste and quality in the production of their own foods, the formalisation of wine tasting externalised the judgment of taste, removing the production of sensory knowledge from producers to become embodied by a group of trained professionals with (almost always) costly education and international experience. Consider, for example, how taste courses I attended featured descriptive language around quality *aszú* wine, with emphasis on notes like "cherimoya" and "passionfruit"; these exotic imports are nearly impossible to locate even in Budapest today, and certainly remain untasted by the vast majority of Tokaji producers.

Liddell reflects on the carelessness with which he observes locals consume wine: without reflection or cognition, without analysis. But as he also notes (47):

> The proper understanding of Hungarian wine culture requires an insight into matters less tangible than laws and research institutes. Wine is in the soul of many Hungarians. It has, for some, an almost sacramental quality. Indeed, when tasting one day, I asked, because I was driving, if I might spit out the samples I was being offered. "Wine," came the reply, "is the blood of God, and to spit it out is sacrilege."

Spitting wine is not uncommon in today's Tokaj, where formalised tastings (and the norms associated with them) are spreading. Nevertheless, many locals continue to consume wine in full, opaque glasses alongside traditional *süti* (baked goods). Dani, who owns a mid-sized winery and guesthouse in Tokaj, took me into his cellar. When I asked him to explain the signage on the door, he laughed, and said it was recovered from a factory nearby where people used to spit tobacco onto the ground. He uses it now to guard his wine cellar and to jokingly remind visitors: "Spitting on the floor is forbidden!"

Policy Landscapes

Simplification: Of Taste, of *Terroir*, of Biodiversity

The political ecology of Tokaj vineyards over the last century solidifies around a move toward simplicity: aiming for legibility (cf. Scott, 1998) in the name of translating *taste*. While labelling schemes such as Geographical Indications (GIs, which include Tokaj's PDO status) are often touted as a way to enhance agrobiodiversity, in this case, it may exacerbate genetic erosion. This is because producers, especially since the 1990s, aim to create *terroir* wines using a single, idealised varietal that now makes up to 80% of new plantings. Such an erosion of agrobiodiversity (relative to archival evidence from the nineteenth and twentieth centuries) is very likely a culprit in recent crop losses due to 'sour' or 'gray' rot.

The turn to furmint and dry winemaking was hinted at in local research of the early 1990s, when there was an expectation that "new market segments [would] appear soon aimed at special consumer habits" (Kecskés and Botos, 1990: 72). The assumption was that consumer habits—and market demands—were varied enough to inspire variations in production. However, the institutional trap of vineyard privatisation, foreign investment, and fragmentation of *dűlő*—coupled with the duality of tastes for 'old'- and 'new'-style, fresh wines—has prevented such direct transformation of the region from one of prescribed, uniform mass production to the variations associated with and prescribed by free enterprise.

Researchers in the early 1990s recognised the need to modernise and adopt a more rigorous controlled system of *dűlő* control to replace the central planning method of communism. In response, the Hungarian wine regions were organised into competitive areas of production and entailed the further simplification of products—and thus, of varietals and production methods—in part, to acquire GI labels. This included the waning of varietals: "So-called world varieties, traditional and recently bred Hungarian varieties are fighting for the leading position in every wine district..." where there is a "high number of varieties in certain districts. This number must be limited in accordance with the character and tradition of the region" (Kecskés and Botos, 1990: 71). Variations in technology and grape-growing within wine regions make it "almost impossible to control them" (72). The same authors prioritise the GI label potential for Hungarian wines, as "wines of this category are more valuable than other wines because the origin, the grape variety, the technology and quality is guaranteed very thoroughly" (72). In Tokaj, quality standards continue to be questioned despite the waning of legal varietals to only six white types, where previously dozens of local red, white, and black types were used (Brawner, 2019). It should be reiterated that this is per official regulations; there remain many local and family producers who grow non-sanctioned types, primarily for local consumption and trade, and wild types occasionally appear at the margins of vineyards (but are normally pulled up); spontaneous crossings are not of concern because vines are reproduced through cloning (Brawner, 2019).

In this insistence that unique Tokaji tastes, along with endemic fungi and native germplasm (Brawner, 2019) have a rightful, place-based identity, it is worth considering the relevance of Vidal's (2005: 48) discussion of agro-nationalism:

> Plenty of myths, all over the world, assume the existence of some sort of exclusive relationship between a particular place and the people who are supposed to have originated from it. But this does not prevent us from realising—whether we like it or not—that migration and displacements of all sorts are really the stuff that history is made of. It would seem however, that whenever it comes to the products of the soil, we seem to lose our sense of historicity.

Throughout history, European nation-crafting has occurred at various scales, including the naturalisation of borders and ethnic groups. A unified and 'authentic' France, for example, was reified as an organic entity through a nature-as-patrimony discourse, while the German nation was predicated on an ethnic ideal (Gangjee, 2012).

The Tokaj *terroir* discussion seems to be caught between the two: the language of the UNESCO HCL nomination cites Hungarian migration, settlement, and cultivation as influenced by "emigrants" and other peripheral outsiders (UNESCO, 2002). At the same time, it presents a unified concept of the Magyar nation as rooted in the Carpathian Basin by way of long-term human-environmental interactions like viticulture. This is nowhere more graphically evident than in the very recent *Bormedence* ('Wine Basin') festival that celebrates "wines, flavors, and experiences from the Carpathian Basin", featuring winemakers from the entire historic territory of Hungary (i.e. 'Greater Hungary') (Figures 2 and 3). Conceived by the 2015 Meeting of Carpathian Basin Winemakers and Musicians (*Kárpát-medencei Borászok és Zenészek Találkozója*), it evokes nostalgia for pre-World War I borders, before Hungary lost two thirds of its land area and a portion of the Tokaj region to Slovakia.

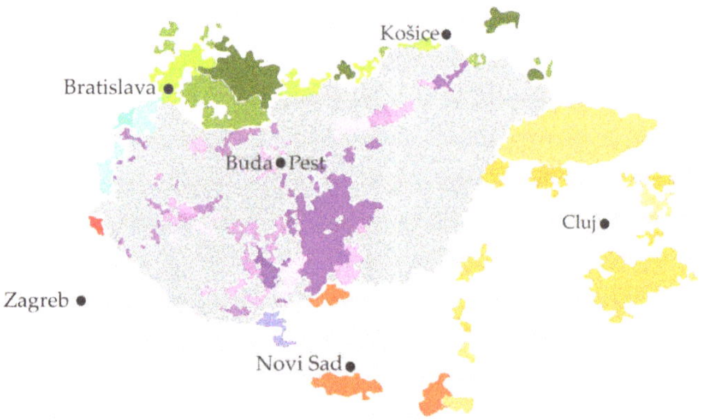

Fig. 2. Map of the territories showcased in the Bormedence wine festival; contemporary Hungary is in dark gray, while the lighter gray area represents pre-Trianon (1920) or "Greater Hungary". Wine regions are located in contemporary Slovakia (green shades), Romania (yellow), Serbia (orange), Croatia (blue), Slovenia (red) and Austria (teal). Image by the author (2021).

The delicate balance between native varietals, the famously fastidious botrytis fungus, and the skilled labour of local people becomes the history of the landscape. In short, through PDOs, claim-making becomes a more-than-human territorial endeavor: native varietals are the "planted flags" (Braverman, 2009) that have marked foreign territories for centuries and reified the Tokaj wine region as Hungarian. Meanwhile, policies themselves become constituent pieces of viticultural ecosystems, material legacies of past ideologies and taste ideals. The commodification of ecologically embedded, or place-based, tastes has

Fig. 3. Greeting arrivals at the Tokaj train station, this image of Greater Hungary (darker, with current Hungary superimposed in lighter stone) reads, "We believe in the resurrection of Hungary!". Photograph by the author (2015).

driven policies that, in turn, materialise in ecosystems through shifting cultivation practices. The legacies of political regimes are made material in the *terroir* itself, which is necessarily somewhat anthropogenic. This suggests the necessity of considering policies (such as the PDO that defines Tokaj) as a part of ecosystems.

Of course, policies emerge as tools of political regimes, requiring us to consider the ways in which political borders interface with *terroir* in this region (e.g. Brawner, 2021; Monterescu, 2017). This chapter focuses on mid-twentieth-century political shifts, however, these must also be put in context. Notably, in October 1918 the Austro-Hungarian dual monarchy was dissolved, to be replaced by a series of (largely unrecognised) interwar republics. On 4 June, 1920, the Treaty of Trianon marked the formal end of World War I and regulated the new independent Hungarian state, including the demarcation of new borders that cost the pre-war Kingdom of Hungary two-thirds of its landed territory. Today, 908.11 hectares of Tokaji *terroir* lies in present-day Slovakia, where *Tokajski* is bottled under the Tokaj PDO label using methods arguably as varied as what can be found in Hungarian Tokaj. While not discussed at length within this chapter, the territory of *terroir* in this context is 'more-than-land'; it is a geography "saturated with national or regional pride, where land-brands are protected as collective intellectual property (e.g. Gangjee, 2012) and world heritage" (Brawner, 2021: 4).

Historic Taste Regimes: The Problem with 'Quality'

Availability of Tokaji wines (and thus, their tastes) have always been politically and geographically contingent. It was "[p]roximity to [Tokaj-] Hegyalja and taste preferences in wine" that "determined the direction of exports of the Tokaji sweet wines during the 16th and 17th centuries" (Lambert-Gocs, 2010: 53). Towards the eighteenth century, mercantilism emerged as an obstacle to Tokaji export going into the nineteenth century as leadership prioritised domestic economies, viewing free trade as a potential threat. This philosophy also gained traction in Poland, perhaps Tokaj's greatest long-standing customer, where "Polish statesmen began having serious doubts about the Polish predilection for Tokaji wine" and the resulting draining of money from domestic products. "This outlook went so far as to envision that Polish tastes could be switched away from grape wine altogether, to the advantage of producers of domestic wines from other fruits of honey" (Lambert-Gocs, 2010: 53).

The Vienna Trade Council convinced the Habsburgs (then under Maria Theresa) to outlaw Hungarian wine export along the Danube and to allow only as much Hungarian wine to be exported as Austrian wine—however, Austrian wine was not in demand, so Hungarian exports were severely limited. Simultaneously, Austria's nationalistic stance promoted their refusal of Prussian goods, and Prussia in turn prohibited Tokaji wine imports. Russia added heavy duties to Hungarian wines in 1766 (with the exception of those purchased for the Russian Imperial Court), giving Tokaji wines in Russia an even more skewed status as the wine of elites.

A simplified story of Tokaji wine history is one in which traditional, sweet "quality" *aszú* wines that once reflected traditional harvesting practices and local ecological circumstances (the endemic botrytis fungus, cellar conditions, etc.) were adulterated by the mode of socialist wine production that overwrote the region in the mid-twentieth century through pressures to produce quantities of artificially sweet wine (seen in opposition to 'quality'). But today, the question of 'quality' wines where GI labels are concerned is simultaneously one of quality *terroir*, where the wines and geographies of western Europe are often cited as benchmarks. Yet, for all its association with communism and 'backwardness', the mid-twentieth-century, quantity-driven production was also simultaneously *en vogue* in post-war western Europe, where places like Burgundy suffered from quantity drives and the "disappointingly thin" wines of the 1970s when the fashion "was to plant clones for quantity and reliability rather than wine quality" (Demossier, 2018: 104). It is therefore important to question the objective "quality" turn that is so often associated with the advent of capitalism in Tokaj. Through these narratives, capitalism and its associated agricultural forms is naturalised through visceral experiences represented as objective through 'good' tastes—while, conversely, the obstinate palates of locals are associated with an equally backwards politics.

Communist producers *were* concerned with quality but were working with internal market demands and Russian tastes (which were, in turn, shaped by production). In fact, communist-era research reports suggest Tokaj had rebounded from an era of low quality following the Napoleonic Wars when exports dropped in the early-nineteenth century and producers resorted to 'lower-quality' production (Bartha, 1974). Additionally, 'foreign' investment (the political boundaries of Central Europe being so mobile) in the region has always played a role, even in the interwar period (the 1920s and early 1930s) when foreign capital, while small, led to the presence of foreign ownership, which incentivised innovation (Csató, 1984). Thus, the use of 'quality' as shorthand for being market-led, and thus associated with the post-1989 era, is oversimplified at best, and seems to stand, instead, alongside broader EU objectives to decrease wine production. The quality in winemaking here is, in other words, in the *exclusivity*. For wine regions that are globally unknown, such as Tokaj, small quantities may be limited and thus exclusive, but do not serve local objectives for export or the building of a place-brand.

Tasting Post-Communist *Terroir*

The socio-political life of taste described here cannot be separated from ecological ramifications. *Aszú* wines made before the industrialisation of communism included stake training (where shoots are tied to a single, often wooden, stake in the ground and shoots climb upward after being horn pruned). Horn pruning creates giant knobs of old, woody bases at the ground with antler-like spurs where new shoots emerge. They would have been originally densely planted (around 10,000 vines per hectare). In the 1970s, Tokaji vine training entailed the Lenz Moser method with five-foot high cordons to assist in the mechanisation of vineyard production; the density of these vineyards would have been determined by the wide rows necessitated by the large Russian tractors used at the time (Liddell, 2001 : 62, Figure 4). They were thus spaced three metres apart, with vines planted at one metre apart at a rate of about 3,330 vines per hectare. As of 2001, new plantations considered 5,000 per hectare to be optimal, with north-south alignment, but most remained closer to 3,330 because of the continued use of old tractors.

Fig. 4. Training vines to cordons, Tarcal, April 2017. Photograph by the author (2017).

The Lenz Moser method is not ideal for varieties prone to rot and may in fact be exacerbating the Gray Rot problems faced in furmint monocultural plantations today, causing reported crop losses of up to 60% (Gabi, personal communication, 2017; corroborated by other respondents). It is typical to get three litres of *must* (grape juice with skins and stems) per vine if growth is restricted; this level of restriction is considered good quality, though some aim for only half of this. However, many locals now sell grapes on to larger corporations rather than making wine themselves. "Peasant growers," writes English wine writer Liddell (2001), "are happy to take all the grapes that God sends, and over-cropping is the Hungarian grape-grower's besetting sin" (62–63). The other "sin", he explains, is "cash-shortage disease", or premature harvesting that leads to underripe grapes being picked in order to sell them (54).

During communism, vines were planted for mass production and responded to cordon planting with vigour, producing as much as twenty tons per hectare; today's top-tier *terroir* wines in Tokaj are harvested at two to four tons per hectare. The grapes were often harvested at marginal ripeness, encountered pasteurisation, fortification, added sugars, or the addition of old wines to hide flaws. These practices were prohibited in the 1991 Wine Act and continue to be associated with supply-side economics and lack of competition. As one journalist wrote of Tokaj in 1990:

> Perversely, the communist regime made life too easy, both for small farmers and the huge co-operative farms. Their crops were already sold, admittedly at low prices, before they were even gathered. Huge yields were therefore all that was required. Selectivity, the first fundamental for wine quality, was a luxury for which only the proudest and most dedicated growers were prepared to pay (Johnson, 1990).

Ironically, the "luxury" of selectivity continues to elude most Tokaj producers today—for lack of market (export) and lack of local demand for the new, fresh taste (and the higher price points associated with low-batch, high-tech production). The Soviet drive to produce bulk wines introduced mechanisation, yet preserved certain traditions, such as barrel ageing. However, it is ostensibly the residue of these steel and iron implements that wine journalists have considered akin to "tasting communism" (Signer, 2015). Reviews of the "unknown *terroir*" of CEE

(Jung, 2014) often cite a tinge of the industrial and are worth a critical eye; notably, and quite in contrast, the adoption of mechanisation in Tuscan vineyards in 2016 was recently lauded as *modernisation* (Ebhardt, 2016).

Fig. 5. Touring a mid-size Tokaji winery, June 2015. Traditional stake training (foreground) requires hefty manual labour (occasionally horse-drawn plows are used), as well as financial investment where terraces are not pre-existing. Photograph by the author (2015).

Today, hilltop vineyards are often viewed as top-tier, because (thanks to the erosion afforded by decades of wide rows during communism), poor soil conditions require the vine to 'work' to reach water and result in lower levels of production (and reportedly higher-quality berries). Communist production would have irrigated these areas or pulled them up entirely (as with many marginal plots). Where vineyards remained un-pulled, rows were cut, and the vines were trained to a high cordon. However, communist practices are to thank for the current conditions (although they rarely enter the *terroir* narrative of producers, and never positively). The 'poor' soil conditions are now thought to train the vine toward quality, while the over-fertilisation of first-class *dűlő* appears

Fig. 6. Surveying the Meleg oldal ("Warm Side") vineyard with Dani, June 2015. Wide spacings between vertical rows, popular since the mid-twentieth century, allow for mechanisation and would require large financial commitments to transform. Photograph by Dan Adams (2015).

to have left legacy nutrients in high quantities. Ironically, this high nutrient content is interpreted by many producers (and consumers!) as the elusive *minerality* that is said to mark quality and volcanic *terroirs* around the world (Brawner, 2018).

Domesticating the Taste of Place: The New 'Authentic'

Tokaj *terroir* is made up of a unique assemblage of geology, ecology, and cultural practices of the region, essentialised and fetishised through endemic wines. As an irreproducible political ecology, the region is protected through international and local policies, including the internationally acknowledged GI. Perhaps the first GI of its kind (with vineyards ranked and enclosed by 1737), Tokaj wine nevertheless depends on a *contemporary* international reputation of quality rooted in its origin (Joslin, 2006). It also relies on the protection of monopoly rents, and demonstrable, regulatable links between geographies and product attributes. Following centuries of political-economic and ecological change, renewed attention to *terroir*-led winemaking in Tokaj aims to restore the region's international standing.

In this chapter, I have argued for the consideration of *terroirs* as political-ecological assemblages. While denaturalised through taste discourses, *terroir* is a site of political ideology and social norms made manifest. In Tokaj, wines have historically been made to suit the values of external markets and 'palates'; what constitutes 'local tradition' may in fact be seen as local responses to external pressures. Further theorisation of these links is required in order to gain insight into the role of human experience and 'ecologically embedded tastes' in political-ecological change. As an ideological geography, *terroir* is an ideal case in which to contemplate the web of politics, social norms, and power relations that underpin coveted wine labels and cast a shadow over the 'unknown' *terroir* of post-socialist Europe (Jung, 2014).

References

Bartha, I., 1974. A Borkereskedés Problémái a Hegyalján a Xix. Század Elso Felében. *Agrartorteneti Szemle*, 16(1/2), 264–76.

Besky, S., 2014. The labor of *terroir* and the *terroir* of labor: Geographical Indication and Darjeeling tea plantations. *Agriculture & Human Values*, 31(1), 83–96. https://doi.org/10.1007/s10460-013-9452-8.

Bodnár, S., 1991. Quality and origin protection of the Tokay wines. Tokaj-Hegyalja State Farm. Sárospatak, Hungary. Vine and Wine Economy, Procedure of the International Symposium. Kecskemét, Hungary, 25–29 June. https://doi.org/10.1016/b978-0-444-98711-2.50009-x.

Brawner, A. J., 2021. Landed value grabbing in the *terroir* of post-socialist specialty wine, *Globalizations*, 18(3), 390–408. https://doi.org/10.1080/14747731.2020.1738101.

——, 2019. *"You can taste it in the wine": A Visceral Political Ecology of Postsocialist Terroir.* The University of Georgia. (Unpublished doctoral thesis). https://getd.libs.uga.edu/pdfs/brawner_amanda_j_201905_phd.pdf.

——, 2018. *The Co-production of Terroir in a Hungarian Wine Region: A Science and Technology Studies (STS) Approach to the Minerality Concept in Viticulture.* The University of Georgia (Unpublished Master's thesis). https://getd.libs.uga.edu/pdfs/brawner_amanda_j_201805_ms.pdf.

Creed, G. W., 1998. *Domesticating Revolution: From Socialist Reform to Ambivalent Transition in a Bulgarian Village*. Pennsylvania State University Press, University Park, PA.

Demossier, M., 2018. *Burgundy: A Global Anthropology of Place and Taste*. Berghahn Books, New York, Oxford. https://doi.org/10.2307/j.ctvw04ffz.8.

Dodds, C., 1991. Cash-strapped Hungarian vineyards appeal to West; Tokaj-Hegyaljai. *The Times*, 21 October.

Fischer, J., Hartel, T., and Kuemmerle, T., 2012. Conservation policy in traditional farming landscapes. *Conservation Letters*, 5(3), 167–75. https://doi.org/10.1111/j.1755-263X.2012.00227.x.

Friedrich, J., 2000. In Hungary, a white-wine renaissance. *The Wall Street Journal*, Western Edition, 5 October, p. 24. https://www.wsj.com/articles/SB970703778890443286.

Gangjee, D., 2012. *Relocating the Law of Geographical Indications*. Cambridge University Press, Cambridge. https://doi.org/10.1017/cbo9781139030939.

Heller, M., 2014. The Commodification of Authenticity, in: Lacoste et al. (Eds.), *Indexing Authenticity: Sociolinguistic Perspectives*. De Gruyter, Berlin, München, Boston, MA, pp. 136–58. https://doi.org/10.1515/9783110347012.136.

Johnson, H., 1990. The finer taste of democracy. *The Sunday Times*, 16 September.

Jung, Y., 2014. Tasting and Judging the Unknown *Terroir* of the Bulgarian Wine: The Political Economy of Sensory Experience. *Food and Foodways*, 22(1–2), 24–47. https://doi.org/10.1080/07409710.2014.892733.

Kovacs, E. K., 2019. Seeing subsidies like a farmer: Emerging subsidy cultures in Hungary. *The Journal of Peasant Studies*, (48)2, 387–410. https://doi.org/10.1080/03066150.2019.1657842.

Krzywoszynska, A., 2015. Wine is not Coca-Cola: marketization and taste in alternative food networks. *Agriculture & Human Values*, 32(3), 491–503. https://doi.org/10.1007/s10460-014-9564-9.

Lambert-Gócs, M., 2010. *Tokaji Wine: Fame, Fate, Tradition*. Wine Appreciation Guild, Limited, California, USA.

Liddell, A., 2003. *The Wines of Hungary (Mitchell Beazley Wine Guides)*. Mitchell Beazley, London.

Monterescu, D., 2017. Border Wines: Terroir across Contested Territory. *Gastronomica: The Journal of Critical Food Studies*, 17(4), 127–40. https://doi.org/10.1525/gfc.2017.17.4.127.

Peet, R., Robbins, P., and Watts, M., 2010. *Global Political Ecology*. Routledge, London. https://doi.org/10.4324/9780203842249.

Signer, R., 2015. Tasting communism in the vineyards of Hungary. *Munchies (Vice)*, 9 November, https://munchies.vice.com/en_us/article/gvmdwy/tasting-communism-in-the-vineyards-of-hungary.

Smith, B. C., 2013. *Questions of Taste. The Philosophy of Wine*. Andrews UK Limited, Luton.

Stahl, J., 2010. *Rent from the Land: A Political Ecology of Postsocialist Rural Transformation*. Anthem Press, London and New York. https://doi.org/10.7135/upo9781843318989.

Tokaji Borvidék Hegyközségi Tanácsa, 2017. A Tokaj oltalom alatt álló eredetmegjelölés termékleírása (8. változat). https://boraszat.kormany.hu/tokaj.

UNESCO, 2020. The 1954 Hague Convention for the Protection of Cultural Property in the Event of Armed Conflict. 14 May. http://www.unesco.org/new/fileadmin/MULTIMEDIA/HQ/CLT/pdf/1954_Convention_EN_2020.pdf.

——, 2002. Tokaj Wine Region Historic Cultural Landscape (World Heritage Nomination), Magyar UNESCO Bizottság, 29 June, https://whc.unesco.org/uploads/nominations/1063.pdf.

Wyatt, C., 2006. Draining France's 'wine lake'. *BBC News*, 10 August, http://news.bbc.co.uk/2/hi/europe/5253006.stm.

10. A Geographical Political Ecology of Eastern European Food Systems

Renata Blumberg

In the midst of debates over the expansion of the European Union's (EU) Common Agricultural Policy (CAP) to future member states in Central and Eastern Europe (CEE), Franz Fischler, the European Commissioner for Agriculture, Rural Development, and Fisheries, stated that introducing the CAP to accession countries "could induce a reluctance to change, hindering the development of sound agricultural structures" (Fischler, 2000). This statement alludes to the power dynamic that shaped the EU accession process for post-socialist nation-states in CEE (Blumberg and Mincyte, 2019; Dzenovska, 2018; Klumbytė, 2011; Kuus, 2004; Smith, 2002). The underlying assumption articulated by Commissioner Fischler, but also circulating in popular discourses, was that existing agricultural structures in this region were not 'sound.' This assumption was based on a deep and persistent association between eastern Europe and 'backwardness' that has long figured in the western European imaginary (Wolff, 1994). This kind of developmentalist thinking has real practical implications because it manifests itself in models, policies and theories that originate in 'advanced' places and are to be applied in 'backward' places.

This chapter uses a geographical political ecology of food systems approach to question this kind of developmentalist thinking, which positions the 'East' as inferior to and needing to learn from the 'West.' It starts with an overview of the agrarian question from the early-twentieth century, wherein eastern Europe figured prominently as a locus of theoretical debate and development. The next section shows how analyses written in and on eastern Europe produced generative

 https://doi.org/10.11647/OBP.0244.10

scholarly insights globally as academic debates were launched about agrarian change and peasant studies with the decline of the European colonies in the twentieth century. Many scholars working in this tradition, which came to be called agrarian political economy, aligned themselves with neo-Leninist Marxism (Bernstein, 2010). As Marxian approaches faced increased critique towards the end of the twentieth century, new theoretical perspectives gained prominence (Buttel, 2001), but the declining influence of Marxism brought about a similar decline in understandings of capitalism in the food system. The end of the section therefore argues for a renewed agrarian political economy approach as part of a broader geographical political ecology of food systems. Building on these insights, the penultimate section takes inspiration from early work on the agrarian question and outlines how a geographical political ecology of food systems in eastern Europe could contribute to broader debates in agrarian and food studies, but also shape geographies of hope, contestation and responsibility in the region. First, it demonstrates the significance of understanding the food system within the context of a 'more-than-capitalist' world. Second, it shows how a geographical political ecology of food systems approach helps explain key developments in food systems in the region. Third, it highlights how scholarly interventions in CEE are pushing theories in agri-food systems in new directions. A geographical political ecology of food systems approach offers the intellectual space for this kind of theoretical development.

The Classic Agrarian Question

In the late-nineteenth century, scholars and activists were confounded by the pace and form of capitalist development in European agriculture. While capitalist industry rapidly transformed urban space and concentrated production at increasingly larger scales, the pace of transformation in rural areas seemed to subside. To the surprise of many, the small-scale peasant or family farm persisted as a dominant organisational form in agriculture in the late-nineteenth century. To socialist revolutionaries, the persistence of peasant agriculture was seen as an impediment to the development of a two-class society of capitalists and workers, which itself was needed to hasten the advance of the communist revolution.

This conundrum prompted detailed study of what became known as the "agrarian question." In his analysis of that subject, Karl Kautsky (1988) found that the persistence of the small farm in the late-nineteenth century did not imply its continued autonomous existence. In fact, Kautsky analysed the manner by which the small (peasant) farm may form a *functional* relationship with large-scale agriculture, a relationship created and spurred on through capitalist development (not in opposition to it, although the presence of peasant farms may superficially suggest that). Various European states had indeed used incentives to promote the establishment of 'undersized' peasant plots and to prevent a potential agricultural labour force from migrating abroad or to the cities (Kautsky, 1988). This arrangement was functional to the extent that the small- and large-scale farms did not compete with each other and that small-scale farms offered a market—and more importantly, a source of labour-power—to large-scale producers. Alternately, according to Kautsky, in the face of competition from large-scale producers, peasant family farms could overexploit their own labour-power. Although 'free peasants' appeared to persist, Kautsky argued that they were increasingly dependent on factories, which had become the only outlet for their production. With the industrialisation and capitalisation of food processing, producer-processing cooperatives struggled to compete with capitalist firms.

Writing about the agrarian question in the Russian context in the early-twentieth century, Lenin (2004 [1899]) undertook a similar analysis of the development of capitalism in agriculture. He found that class-based differentiation between peasants was already taking place: the peasantry had ceased to exist as a (feudal) class and was constituted by internally differentiated positions of the rural bourgeoisie and the rural proletariat. Although a few of the remaining 'middle peasants' would join the bourgeoisie, Lenin predicted that they would generally be flung into the masses of the rural proletariat by undergoing a process of de-peasantisation. The rural bourgeoisie were defined by their employment through wage labour and their possession of larger holdings. The Russian rural proletariat, in contrast, typically farmed on small allotments. Lenin maintained that this appearance of 'peasant' production was misleading and underscored the fact that agricultural holdings did not preclude dependence on wage labour for survival. He also analysed different paths of capitalist development: for example, the American path, led by free peasants; and the Prussian path, led by the landed nobility with their large estates (Bernstein, 1996). He maintained that although capitalist transitions occurred differently over space and manifested themselves in different forms, the ultimate result everywhere would be differentiation into two classes: bourgeoisie and proletariat. In addition to Lenin's theoretical analyses on the Russian context, Alexander Chayanov offered another influential view on the Soviet peasant economy in the 1920s. Before forced collectivisation, Chayanov and other agrarian economists (or social agronomists) set out to thoroughly study the peasant household, even if such a study entailed bracketing and isolating certain phenomena for the purposes of theoretical abstraction (1986). Although his views evolved in other work (1991), Chayanov showed how social differentiation in the countryside was a function of demographic change in the life-cycle of the household. He also attempted to distil the unconscious logic that propelled the peasant household economy in times of crisis as well as in times of abundance. Like Kautsky, Chayanov argued that the family farm's competitive power is fueled by self-exploitation, or the capacity of peasant families to work more (and harder) in order to satisfy their needs. Therefore, in an economy of declining prices for agricultural goods, capitalist firms have to cut back on production, but peasant farms actually work more to make enough income (and thus maximise total income, not profit).

Faced with the growing militancy of peasant movements in the Third World in the 1960s, including the rising prominence of Maoism, scholars of development studies sought to apply the insights of Lenin, Chayanov, and others on agrarian transitions to capitalism outside of Europe and Soviet-controlled territories. Challenged by the realities of different socio-spatial contexts, scholars refined concepts to account for the complexities they encountered and the scholarly trajectory around the field of agrarian political economy grew. In the following section I discuss some of these contributions, highlight subsequent critiques of agrarian political economy, and formulate a new approach based in political ecology that draws upon the insights of agrarian political economy while addressing some of its criticisms. I demonstrate how the insights of these early scholars of the agrarian question continue to have relevance today by providing a way to understand processes such as de/re-peasantisation taking place throughout CEE.

From Agrarian Political Economy to a Geographical Political Ecology of Food Systems

The writings of Lenin and Chayanov provided the theoretical foundation for a generative field of social research on agrarian transitions, which has produced insights globally. In this section I highlight some of these contributions and their associated debates before going on to detail some internal and external critiques of agrarian political economy. I conclude the section by describing a geographical political ecology of food systems approach, which includes a renewed agrarian political economy that accounts for these critiques.

Agrarian political economists writing in Asia (Akram-Lodhi, 2005; Zhang, 2015), Africa (Bernstein, 1979; Levin and Neocosmos, 1989), and Latin America (de Janvry, 1981; Kay, 1981) who based their work on Lenin's insights, argued that farmer livelihoods are bound up within the dynamics of capitalism, producing a changing and differentiated social landscape (see Bernstein and Byres, 2001 for a more comprehensive analysis of this literature). Class analysis yielded meaningful insights into this differentiated social landscape and aided in the recognition, often against the claims of farmers' movements, of farmers' disparate interests. Even the classification of a 'small' or 'mid-sized' farmer is an analytically weak one that says nothing about the position of that farmer in a socially, economically and ecologically differentiated world (Bernstein, 2010). Following Lenin, who pointed out that the transitions from feudal class relations to capitalist relations vary across space, scholars have also documented the multiple paths that agrarian transitions have taken (Bernstein, 2010). What is common in all of them is the commodification of subsistence, which may not be total (i.e. labour may not be commodified), but which still forces peasants to depend in some manner or another on commodity relations for subsistence. Finally, this research also recognised the significance of studying existing patterns of capital accumulation in a structurally heterogeneous world economic system, a system that produces differences between central and peripheral nation-states (de Janvry, 1981). In other words, conditions labelled as 'backward' came to be understood as produced thus through uneven development.

Many scholars sought alternative explanations to make sense of peasant responses to the expansion of capitalism by applying Chayanov's theories. These scholars argued that peasant production constituted a unique mode of production within capitalism, one governed by its own logic (Vergopoulos, 1978), putting them at odds with others who drew more directly from Lenin. In the subsequent so-called Lenin-Chayanov debate (see Banaji, 1976; Bernstein and Byres, 2001; Bernstein, 2009), Lenin and Chayanov were cast as theoretical adversaries (Bernstein, 2009). Other scholars sought to reconcile Chayanovian and Marxist approaches. In her history of simple commodity production, Friedmann's approach (1978), since labeled Chayanovian Marxism (Buttel, 2001), argued against the predominant assumption that simple commodity producers will always lose out to capitalist producers.

Although Chayanov assumed an ontology that was different from that developed by Lenin, what binds the work of Lenin and Chayanov is their common concern with the social relations of production (capitalist or non-capitalist) and their privileging of an economistic understanding of agrarian transitions and change. While the focus on the relationship between class dynamics and agrarian change and more broadly on the social relations of production remained dominant in peasant studies and agrarian political economy in the mid- to late-twentieth century, critiques began to arise about the narrowness of this focus. In particular, feminist scholars drew attention to the significance of patriarchal relations in the household and to gender relations more broadly in the food system (Razavi, 2009; Ramamurthy, 2000).

Other scholars criticised the focus on the social structure of agriculture and argued for the need to understand the relationship between nature and agriculture. For example, Goodman, Sorj, and Wilkinson (1987) reconsidered the role that nature plays in preventing the transformation of agricultural production into a unified industrial process. Farming systems are based in biophysical processes, which shape and constrain production. For example, weather, pests, and diseases may exact a considerable and unpredictable toll on production, which itself cannot be fully controlled because of the seasonality of agricultural production. However, the authors argue that industrial capital has been able to adapt to natural constraints by employing strategies of "substitutionism", accumulation from the processing of agricultural outputs, and "appropriationism", when value-generating activities move out of the direct sphere of the farmer, effectively commodifying farm processes.

By the end of the twentieth century, the new global reality of agro-food systems had become a major focus of research in agrarian political economy (Watts and Goodman, 1997). For example, the concept of "food regime" was created to historicise the global political economy of food by attempting to account for the multiple factors that contribute to periods of stability, transition and crisis in capital accumulation (see McMichael, 2009 for a genealogy of food regimes). Studies also focused on the growing power of transnational corporations (TNCs), their organisational and operational structures, and their relationships with nation-states and other governing bodies (Bonanno et al., 1994). The rise of TNCs went hand in hand with the growing dominance of neoliberalism (Watts and Goodman, 1997).

Despite efforts to shift the focus of agrarian political economy to issues beyond the social relations of production, in the last decade of the twentieth century, the theoretical influence of agrarian political economy declined (Buttel, 2001). Scholars increasingly critiqued agrarian political economy as a theoretical perspective, arguing that it overlooked the agency of nature, did not adequately theorise consumption and culture, and neglected peasant agency as potentially transformative and politically meaningful (Buttel, 2001). According to some critics, by focusing exclusively on capitalist transformations, existing and past scholarship had privileged certain economic practices over others and therefore neglected to adequately consider the importance of a wide spectrum of non-capitalist practices (Gibson-Graham, 1996; Whatmore and Thorne, 1997). Critics of 'capitalocentrism' have further argued that an exclusive focus on global capitalism and conventional food chains deflects attention from already existing alternatives, which could otherwise be recognised, strengthened and sustained (Whatmore and Thorne, 1997). Furthermore, because discourses are themselves productive of the worlds they seek to represent, scholarship that represents food systems as globalised and exclusively capitalist helps to produce a world in which seeing and supporting alternative food systems becomes more difficult. To call attention to alternatives and to thereby help shape a world that fosters alternative food systems, Whatmore and Thorne (1997) suggest an approach in which they are made visible and meaningful.

Scholarship in agrarian political economy has also been accused of applying a restrictive understanding of nature and materiality (Bakker and Bridge, 2006). Even when scholars actively took an interest in nature, it was represented as an obstacle or constraint (Goodman, Sorj, and Wilkinson, 1987; Mann, 1990), or as a vehicle for capital accumulation (Boyd, Prudham, and Schurman, 2001). This critique has extended to Marxist approaches more broadly, which have been criticised not only for restricted understandings of nature but for anthropocentrism and for reproducing problematic dualistic divides (Castree, 2002).

The focus on production in agrarian political economy reflects the assumption by Marxists that the sphere of production is the sole locus of political agency and potential transformative power. This is where labour meets capital, where surplus value is extracted from workers through the labour process. Consumption, merely a component of the sphere of exchange, has been generally relegated to a lesser, or even invisible, position. While consumer movements driven by environmental and other concerns made their presence felt in the food chain in the late-twentieth century, scholars had not developed concepts that could provide a more nuanced account of a kind of consumption that was clearly "more than merely a niche marketing opportunity" (Goodman and Dupuis, 2002: 18). For consumers, who were otherwise absent from the productive locus of power, political engagement could only be pursued through the unveiling of commodity fetishism. Consumers were not the only agents lacking transformative power according to agrarian political economists. The insistence that peasants did not form a distinct class (neither completely capitalist, nor completely proletariat) led scholars to dismiss or criticise movements that rallied on behalf of peasants. However, peasant movements remained prominent globally, even exerting political agency (Edelman, 1999).

To explain consumer and peasant movements, and to seriously evaluate possible alternatives to the conventional food system, scholars began to use multiple theoretical perspectives, from convention theory to post-humanist approaches (Blumberg et al., 2020; Goodman et al., 2012; Le Velly and Dufeu, 2016; Murdoch et al., 2000; Ponte, 2016; Wills and

Arundel, 2017). These theoretical approaches have become especially prominent in the study of 'alternative food networks,' which encompass the direct-to-consumer marketing outlets, such as farmers' markets, which have grown in recent years (Blumberg et al., 2020; Goodman et al., 2012). Research on the broader alternative geographies of food, from farmers' markets to urban agriculture, has produced significant insights on the strengths and limitations of these initiatives. To fully understand the source of these limitations, some scholars have insisted on the on-going relevance of political economy approaches, including agrarian political economy. For example, in her study on organic farming in California, Guthman (2004) demonstrated that capitalist dynamics were transforming the organic sector in a way that was compromising some organic ideals. Similarly, Galt (2013b) drew upon agrarian political economy to analyse farmer earnings in community supported agriculture, and he found high rates of self-exploitation. Despite the ubiquity of alternative geographies of food, capitalist dynamics have continued to influence the food system, from the proliferation of unhealthy dietary patterns (Otero, 2018), to the global expansion of land grabbing (Hall, 2013), to the more recent devastation unleashed by the coronavirus pandemic and other emergent diseases (Wallace et al., 2016; Wallace, 2020).

Clearly, despite the presence of alterity in the food system, it remains important to understand how capitalist dynamics influence the sustainability of food systems. As a result, Galt (2013b) has argued for the continued relevance of agrarian political economy, but he also acknowledges its weaknesses (Galt 2013a). To address these shortcomings, Galt (2013a) draws inspiration from the field of political ecology, and he argues that insights from agrarian political economy could be harnessed and integrated into a 'political ecology of agrifood systems.' By focusing on everyday lived environments, the politics of food production, struggles over the commons and many other related topics, political ecology has drawn connections between power relations and environmental change (Galt, 2013a).

While much political ecology research has focused on the local scale, it is not inimical to multi-scalar analysis (Engel-Di Mauro, 2009). Building on Galt's (2013a) initial model, Blumberg et al. (2020) argue for the need to integrate a geographical approach to the political ecology of food systems through spatial concepts, such as Massey's understanding of space as a heterogeneous multiplicity (2007). Understanding space as a multiplicity of processes, and thus as always in process, constituting and being constituted by flows, problematises developmentalist thinking that positions some places as 'behind' others, just as eastern Europe has often been cast as 'behind' western Europe. Space is also laden with power-geometries because the multiplicity of trajectories that constitute space are not equal in their capacities. As Massey explains, "understanding space as the constant open production of the topologies of power points to the fact that different 'places' will stand in contrasting relations to the global. They are differentially located in wider power-geometries" (Massey, 2005: 101). In their conceptualisation of a geographical political ecology of food systems framework, Blumberg et al. (2020) draw upon Massey's conception of space as a heterogeneous multiplicity to demonstrate how capitalist dynamics intermingle with other trajectories to make space. However, any use of agrarian political economy must address limitations surrounding its core ontological foundations.

Agrarian political economy has been focused on class analysis and differentiation because in Marxist theory the proletariat plays an important role in overcoming capitalism. "As is well known, for Marx the possibility of transcending capitalism lay in the hands of the class that it created: only the proletariat, a class free from the ownership of the means of production and free to sell its labour-power, was capable of eradicating class society and ending exploitation" (Akram-Lodhi and Kay, 2010: 181). In this manner, agrarian political economy resembles what Moishe Postone calls 'traditional Marxism,' which includes "theoretical approaches that analyze capitalism from the standpoint of labor and characterize that society essentially in terms of class relations, structured by private ownership of the means of production and a market-regulated economy" (1993: 7).

A significant amount of scholarship has been devoted to critically assessing traditional Marxism, especially by heterodox Marxists who remain committed to the critique of capitalism, but understand the necessity of providing alternative conceptualisations of its overcoming (Clough and Blumberg, 2012; Postone, 1993). Socialist and Marxist feminists in particular have demonstrated the significance of social reproduction as an analytical site to analyse how capitalism depends on social reproduction even as it destroys its basis (Federici, 2012; Katz, 2001). Like traditional Marxism, agrarian political economy has privileged the study of class dynamics, without taking seriously the forms of domination that have emerged in capitalist society. A shift away from 'traditional' Marxism, and in my application away from 'traditional' toward a 'renewed' agrarian political economy, does not discard all of traditional Marxism's insights. Instead, it situates these insights within an alternative, but still Marxist perspective. Capitalism remains a contradictory, crisis-ridden system, prone to overproduction and underproduction. A renewed agrarian political economy allows us to account for consumer and peasant agency, to take alterity and nature seriously and to understand the importance of capitalism without relying on reductionism.

In the contemporary context, more and more consumers are concerned about food safety, and they seek to secure healthy, organic food. Likewise, more producers are willing to grow this food. However, "what characterizes capitalism is that, on a deep systemic level, production is not for the sake of consumption. Rather, it is driven, ultimately, by a system of abstract compulsions constituted by the double character of labor in capitalism, which posit production as its own goal" (Postone, 1993: 184). These abstract compulsions that cause producers/farmers to intensify production result in only short-term increases in surplus value generated. Once increases in productivity become socially general, the value generated per unit decreases. Even alternative agricultural production is only partially driven by the needs or desires of the local consumer; ultimately, it is driven by production for accumulation.

Drawing upon Moishe Postone's work, Noel Castree (1999: 141) argues that: "capitalism can be seen as a constitutively 'open' system which, while structured, global and hegemonic, is nonetheless constantly infused by its putatively 'non-capitalist' exteriors." In short, capitalist development does not cancel other logics and one can conceptualise a systemic capitalism without reducing all logics to capitalist logic. In this system, non-commodified and commodified labour may exist side-by-side: there is no linear progression to more and more commodified labour, or fewer and fewer peasant or household producers. Indeed, a geographical political ecology of food systems approach underscores that circuits of capital structure practices, but they do not homogenise concrete labours or space.

In the following section I apply a geographical political ecology of food systems approach in an analysis of eastern European food systems since the late-twentieth century. The goal is not to provide a comprehensive overview of the multiple and complex changes that have occurred in food systems throughout the region. Instead, I connect these changes with the consolidation of neoliberal globalisation in the 1990s and the production of uneven geographical development and social inequality, both of which are important for understanding the geographical political ecology of food systems and have long been concerns within agrarian political economy. Harking back to Chayanov (1986), I also recognise the existence of multiple logics, some of which are making pathways towards more sustainable development in the region. Finally, I demonstrate how and why a geographical political ecology of food systems approach is useful in understanding both the possibilities and limitations inherent in these alternative food geographies.

Geographical Political Ecology of Eastern European Food Systems

Eastern European food systems have undergone radical transformations since the collapse of state-socialism in the late-twentieth century, and these changes exemplify both the expansion and consolidation of neoliberal globalisation, as well as the complex shaping and reshaping of non-capitalist logics, including those governing subsistence food production. Neoliberal policies implemented through shock therapy in the 1990s, the expansion of the European Union, and austerity policies following the 2008/09 financial crisis have transformed food production and consumption (Stuckler and Basu, 2013; Caldwell, 2009; Jung et al., 2014; Woolfson and Sommers, 2016), and rural geographies more broadly (Bohle and Greskovits, 2007; Burawoy et al., 2000; Burawoy and Verdery, 1999; Creed, 1998; Leonard and Kaneff, 2002; Verdery, 2003). As a consequence of the European Union's adherence to neoliberal policies and global competitiveness, uneven geographical development has also reshaped the landscape (Agnew, 2001; Dunford and Smith, 2000; Hudson, 2003).

Since the 1990s, the agricultural sector was negatively affected by cheaper, subsidised food imported from western Europe (Blumberg and Mincyte, 2019; Engel-Di Mauro, 2006). Indeed, the power-geometries that span space, shape capitalist development in profound ways. For example, writing about the Polish sugar beet industry, Kim (2011; 2012) documented the negative effects of increasing competition and privatisation: in an effort to attract foreign investment and modernise plants, factories were sold to corporations based in western Europe. Many of these factories were later closed by those corporations, in part because of new EU policies that provided compensation for limiting sugar production (Kim, 2012). Farmers were adversely affected, while also facing rising costs and fluctuating prices.

Faced with these difficulties in conventional supply chains, many farmers have turned to supply consumers directly through local or alternative food networks. This phenomenon has been growing throughout CEE (Balázs et al., 2016; Benedek and Balázs, 2016; Bilewicz and Śpiewak, 2018; Grivins and Tisenkopfs, 2015; Mincyte, 2012; Smeds, 2015; Spilková et al., 2013; Spilková and Perlín, 2013; Syrovátková, 2016; Syrovátková et al., 2015). Nevertheless, alternative food networks are still shaped by competition, which is rarely conceptualised and theorised in the literature on alternative food networks. In a case study in Lithuania, Blumberg (2015; 2018) documented how a dramatic fall in milk prices and the global financial crisis both enhanced interest in local food among consumers and enhanced competition for farmers seeking to provide that food to the local market. As a result, the promotion of alternative food networks has only furthered differentiation between farmers, as many mid- and large-scale farmers were able to take long-term advantage of these opportunities (Blumberg, 2018).

While the dynamics of capitalist accumulation help explain these transformations, they cannot fully account for the way political, social and cultural geographies also shaped these processes (Aistara, 2011; Kovács, 2015; Schwartz, 2005, 2007). By integrating an understanding of the more-than-capitalist world, a geographical political ecology of food systems approach helps explain key developments in food systems in the region. For example, longstanding practices of subsistence food production are part of the social geographies that have helped households manage social reproduction during times of dramatic changes (Mincyte, 2011; Mincytė et al., 2020). In the 1990s, when Polish rural communities were beginning to feel the impacts of market integration, they responded by relying more heavily on what appeared to be older social arrangements of producing food, such as intensified subsistence peasant production (Zbierski-Salameh, 1999). Although certain social practices may bear resemblance to older forms, in post-socialism, their causes were actually novel (Burawoy and Verdery, 1999). Rather than being signs of 'backwardness,' or relics of the past, they are partially products of expansionary capitalism characterised by social domination, increasingly fragmented labour and globalised agri-food networks that have enforced rationalisation and industrialisation of the production process. These processes of differentiation were accelerated and promoted through the application of the EU's CAP and stringent food safety legislation.

The financial crisis of 2008 and the subsequent implementation of austerity policies throughout the Central and Eastern European region also had a profound impact on consumers. As in the past, economic hardships encouraged practices like subsistence food production and exchange as part of larger informal economies (Blumberg and Mincyte, 2019; Smith et al., 2008; Staddon, 2009; Smith and Stenning, 2006). While explanations for the persistence of household self-provisioning have long featured in academic literature (Czegledy, 2002; Hann, 2003; Seeth et al., 1998; Humphrey and Mandel, 2002), only recently have scholars considered how alternative food production, procurement and marketing practices provide possible sustainable development pathways (Ančić et al., 2019; Blumberg and Mincyte, 2019; Blumberg, 2018; Brawner, 2015; Jehlička, 2021; Jehlička et al., 2020; Pungas, 2019; Smith and Jehlička, 2013; Spilková and Vágner, 2018; Yotova, 2018).

The ubiquity and complexity of alternative food practices in the region is pushing scholarship on alternative food networks in new directions. Rather than assuming the universality of concepts such as 'local food,' 'alternative food networks,' or 'farmers' markets,' scholars have documented how concepts travel, merge and are transformed in local contexts (Bilewicz, 2020). These concepts are part of the trajectories that make space (Massey, 2005). For example, Goszczyński and colleagues (2019) highlight how concepts related to alternative food networks have travelled and been adopted and adapted in Poland, while Fendrychová and Jehlička (2018) examine the travelling concept of a farmers' market in Czechia. In both cases, the authors show how meaning and understanding cannot be presupposed, and how the historical geographies of food provisioning and marketing influence and transform travelling concepts. Writing about these historical geographies in CEE, Goszczyński and colleagues (2019) propose the concept of 'invisible alternativeness' to capture the fact that everyday and embedded non-industrial food production, distribution and consumption practices have the potential to remake food regimes, even though they may not be viewed as unique or alternative locally. Similarly, in their study of food self-provisioning in the Czech Republic, Jehlička and colleagues (2019) demonstrate that food self-provisioning can be considered as a form of resilience that counters neoliberalism and enables transformations in the food system.

While this scholarship does not explicitly engage with political ecology for the most part, a geographical political ecology of food systems approach offers the intellectual milieu for broader theoretical development because it weaves together political ecology, geography and agrarian political economy. All three fields combined under this framework are particularly useful for understanding food systems in CEE. For example, political ecology research has long been concerned with the politics of the commons, especially by critiquing efforts of enclosure, whether they be led by the state or private entities (Turner, 2017). In CEE, an important example of 'actually existing commons' (Turner, 2017) are the forests that provide abundant resources for foraging throughout the region. Forest resources, such as berries and mushrooms, enhance food security (Łuczaj et al., 2012), provide livelihood opportunities (Sõukand et al., 2020) and serve as a reservoir to maintain ecological knowledge (Pieroni and Sõukand, 2018). However, they can also be over-exploited and impacted by the expansion of industrial agriculture practices (Łuczaj et al., 2012). Agrarian political economy provides the tools to examine capitalist accumulation in the food system, and how it manifests through the industrialisation of agriculture and differentiation among producers. Writing about pork producers in Poland, Mroczkowska (2019) documents the process of differentiation, and connects it to the multi-scalar politics of EU agricultural policy. She finds that small-scale pork producers have been marginalised from the formal economic sphere, yet the production and consumption of pork by these farmers continues and carries a moral distinction associated with *swoje/swojskie* (our, familiar) food, as opposed to store-bought food (Mroczkowska, 2019). Geographic research on the politics of scale provides tools to understand the formation and implementation of policy in the food system, and its unintended effects (Blumberg and Mincyte, 2020). As calls for a new kind of food policy in Europe intensify (De Schutter et al., 2020), this kind of research is especially significant.

Conclusion

On a beautiful August day in 2010, I spent the afternoon sitting in one of Riga's many cafes discussing the difficulties and possibilities in establishing formalised alternative food networks with Ieva (a pseudonym), a woman who was in the midst of organising a collective direct marketing initiative. Ieva's alternative food network was inspired by the community-supported agriculture (CSA) systems that exist in other countries, but it differed from them in at least two respects: it did not require pre-payment for the whole season and it included a handful of organic farmers who offered different products. Ieva's system was basically a collective purchasing cooperative in which consumers would place orders on a weekly basis, farmers would cooperate to deliver the products to a set location in Riga, and consumers would volunteer on a rotating basis to process the orders. The financial constraints of the participating consumers influenced the structure of the network. A CSA model with pre-paid seasonal shares was not possible because most participating consumers would not have the required money in advance. Furthermore, the idea of receiving items that were not specifically chosen (as usually occurs in CSA or box schemes) was not appealing because the consumers involved already spend at least 30% of their household income on food and they did not want to waste money purchasing unwanted items.

Within a year, Ieva's consumer group was operating smoothly and interest had grown so much that Ieva's group had to turn prospective consumers away. However, she did help other groups to get started.

They are able to order organic and locally-grown food for prices that are generally lower than those in stores. Collective purchasing has also fostered a sense of community among the consumers and a feeling of reconnection with producers. For participating farmers, these initiatives have brought benefits too: farmers get paid immediately and with orders in place, and they know exactly what to deliver. As a result, farmers generally welcome these new initiatives, but they have no illusion that they can exclusively rely on them for their livelihoods; thus far only a small portion of participating farmers' total sales are made through them. Ordering also becomes more erratic in the summer months when people take vacations. This is precisely the time when farmers experience a glut in available produce. Therefore, farmers must still seek out other markets or channels, such as export markets or conventional supply chains, in order to sustain their livelihoods.

By maintaining an understanding of capitalist dynamics without reducing all phenomena to capitalism, a geographical political ecology of food systems approach helps explain the limitations of models like CSA. Through uneven geographical development, eastern Europe continues to experience higher rates of poverty and food insecurity (Davis and Geiger, 2017; Garratt, 2020). Despite consumer interest, CSAs are not ubiquitous because of their dependence on models which assume high disposable incomes. Nevertheless, these systems and other invisible alternatives (Goszczyński et al., 2019) continue to be practiced, remade and adapted.

In the twentieth century, scholars of agrarian change were heavily influenced by theoretical research on agrarian transitions in eastern Europe. While the influence of this theoretical research and the scholarship it generated has waned in the broader study of agri-food systems, it produced important insights that remain relevant in explaining problems in the political ecology of food systems, as well as pointing towards solutions. Using an approach based on a renewed agrarian political economy, which is part of a geographical political ecology of food systems, this chapter draws attention to new strands of research in the region that are once again pushing theories in new directions.

References

Agnew, J., 2001. How Many Europes?: The European Union, Eastward Enlargement and Uneven Development. *European Urban and Regional Studies*, 8(1), 29–38. https://doi.org/10.1177/096977640100800103.

Aistara, G. A., 2011. Seeds of Kin, Kin of Seeds: The Commodification of Organic Seeds and Social Relations in Costa Rica and Latvia. *Ethnography*, 12(4), 490–517. https://doi.org/10.1177/1466138111400721.

Akram-Lodhi, H. A., and Kay, C., 2010. Surveying the Agrarian Question (Part 1): Unearthing Foundations, Exploring Diversity. *The Journal of Peasant Studies*, 37(1), 177–202. https://doi.org/10.1080/03066150903498838.

Akram-Lodhi, H. A., 2005. Vietnam's Agriculture: Processes of Rich Peasant Accumulation and Mechanisms of Social Differentiation. *Journal of Agrarian Change*, 5(1), 73–116. https://doi.org/10.1111/j.1471-0366.2004.00095.x.

Ančić, B., Domazet, M., and Župarić-Iljić, D., 2019. 'For My Health and for My Friends': Exploring Motivation, Sharing, Environmentalism, Resilience and Class Structure of Food Self-Provisioning. *Geoforum*, 106, 68–77. https://doi.org/10.1016/j.geoforum.2019.07.018.

Bakker, K., and Bridge, G., 2006. Material Worlds? Resource Geographies and the 'Matter of Nature.' *Progress in Human Geography*, 30(1), 5–27. https://doi.org/10.1191/0309132506ph588oa.

Balázs, B., Pataki, G., and Lazányi, O., 2016. Prospects for the Future: Community Supported Agriculture in Hungary. *Futures*, 83, 100–11. https://doi.org/10.1016/j.futures.2016.03.005.

Banaji, J., 1976. Chayanov, Kautsky, Lenin: Considerations towards a Synthesis. *Economic and Political Weekly*, 11(40), 1594–607. https://doi.org/10.2307/4364979.

Benedek, Z., and Balázs, B., 2016. Current Status and Future Prospect of Local Food Production in Hungary: A Spatial Analysis. *European Planning Studies*, 24(3), 607–24. https://doi.org/10.1080/09654313.2015.1096325.

Bernstein, H., 2010. *Class Dynamics of Agrarian Change*. Kumarian Press, Sterling, VA.

——, 2009. V. I. Lenin and A. V. Chayanov: Looking Back, Looking Forward. *The Journal of Peasant Studies*, 36(1), 55–81. https://doi.org/10.1080/03066150902820289.

——, 1996. Agrarian Questions Then and Now. *The Journal of Peasant Studies*, 24(1–2), 22–59. https://doi.org/10.1080/03066159608438630.

——, 1979. African Peasantries: A Theoretical Framework. *The Journal of Peasant Studies*, 6(4), 421–43. https://doi.org/10.1080/03066157908438084.

Bernstein, H., and Byres, T., 2001. From Peasant Studies to Agrarian Change. *Journal of Agrarian Change*, 1(1), 1–56. https://doi.org/10.1111/1471-0366.00002.

Bilewicz, A. M., 2020. Beyond the Modernisation Paradigm: Elements of a Food Sovereignty Discourse in Farmer Protest Movements and Alternative Food Networks in Poland. *Sociologia Ruralis*, 60(4), 754–72. https://doi.org/10.1111/soru.12295.

Bilewicz, A., and Śpiewak, R., 2018. Beyond the 'Northern' and 'Southern' Divide: Food and Space in Polish Consumer Cooperatives. *East European Politics and Societies: And Cultures*, 33(3), 579–602. https://doi.org/10.1177/0888325418806046.

Blumberg, R., 2018. Alternative Food Networks and Farmer Livelihoods: A Spatializing Livelihoods Perspective. *Geoforum*, 88, 161–73. https://doi.org/10.1016/j.geoforum.2017.10.007.

——, 2015. Geographies of Reconnection at the Marketplace. *Journal of Baltic Studies*, 46(3), 299–318. https://doi.org/10.1080/01629778.2015.1073917.

Blumberg, R., and Mincyte, D., 2020. Beyond Europeanization: The Politics of Scale and Positionality in Lithuania's Alternative Food Networks. *European Urban and Regional Studies*, 27(2), 189–205. https://doi.org/10.1177/0969776419881174.

——, 2019. Infrastructures of Taste: Rethinking Local Food Histories in Lithuania. *Appetite* 138, 252–59. https://doi.org/10.1016/j.appet.2019.02.016.

Blumberg, R., Leitner, H., and Cadieux, K. V., 2020. For Food Space: Theorizing Alternative Food Networks Beyond Alterity. *Journal of Political Ecology*. 27(1), 1–22. https://doi.org/10.2458/v27i1.23026.

Bohle, D., and Greskovits, B., 2007. Neoliberalism, Embedded Neoliberalism and Neocorporatism: Towards Transnational Capitalism in Central-Eastern Europe. *West European Politics*, 30(3), 443–66. https://doi.org/10.1080/01402380701276287.

Bonanno, A., Busch, L., Friedland, W. H., Gouveia, L., and Mingione, E. (Eds.), 1994. *From Columbus to ConAgra: The Globalization of Agriculture and Food*. University Press of Kansas, Lawrence, KS.

Boyd, W., Prudham, W. S., and Schurman, R. A., 2001. Industrial Dynamics and the Problem of Nature. *Society & Natural Resources*, 14(7), 555–70. https://doi.org/10.1080/08941920120686.

Brawner, A. J., 2015. Permaculture in the Margins: Realizing Central European Regeneration. *Journal of Political Ecology*, 22(1), 429–44. https://doi.org/10.2458/v22i1.21117.

Burawoy, M., and Verdery, K. (Eds.), 1999. *Uncertain Transition: Ethnographies of Change in the Postsocialist World*. Rowman & Littlefield, Oxford.

Burawoy, M., Krotov, P., and Lytkina, T., 2000. *Involution and Destitution in Capitalist Russia. Ethnography*, 1(1), 43–65. https://doi.org/10.1177/14661380022230633.

Buttel, F. H., 2001. Some Reflections on Late Twentieth Century Agrarian Political Economy. *Sociologia Ruralis*, 41(2), 165–81. https://doi.org/10.1111/1467-9523.00176.

Caldwell, M. L. (Ed.), 2009. *Food & Everyday Life in the Postsocialist World*. Indiana University Press, Bloomingdale, IN.

Castree, N., 2002. False Antitheses? Marxism, Nature and Actor-Networks. *Antipode*, 34(1), 111–46. https://doi.org/10.1111/1467-8330.00228.

——, 1999. Envisioning Capitalism: Geography and the Renewal of Marxian Political Economy. *Transactions of the Institute of British Geographers*, 24(2), 137–58. https://doi.org/10.1111/j.0020-2754.1999.00137.x.

Chayanov, A., 1991. *The Theory of Peasant Co-Operatives*. Ohio State University Press, Columbus, OH.

——, 1986. *Theory of Peasant Economy*. University of Wisconsin Press, Madison, WI.

Clough, N., and Blumberg, R., 2012. Toward Anarchist and Autonomist Marxist Geographies. *ACME: An International Journal for Critical Geographies*, 11(3), 335–51.

Creed, G. W., 1998. *Domesticating Revolution: From Socialist Reform to Ambivalent Transition in a Bulgarian Village*. Pennsylvania State University Press, University Park, PA.

Czegledy, A., 2002. Urban Peasants in a Post-Socialist World: Small-Scale Agriculturalists in Hungary, in: Leonard, P., and Kaneff, D. (Eds), *Post-Socialist Peasant? Rural and Urban Constructions of Identity in Eastern Europe*, pp. 200–20. Palgrave, London.

Davis, O., and Baumberg Geiger, B., 2017. Did Food Insecurity rise across Europe after the 2008 Crisis? An analysis across welfare regimes. *Social Policy and Society*, 16(3), 343–60. https://doi.org/10.1017/s1474746416000166.

De Janvry, A., 1981. *The Agrarian Question and Reformism in Latin America*. Johns Hopkins University Press, Baltimore, MA.

De Schutter, O., Jacobs, N., and Clément, C., 2020. A 'Common Food Policy' for Europe: How Governance Reforms Can Spark a Shift to Healthy Diets and Sustainable Food Systems. *Food Policy*, 96. https://doi.org/10.1016/j.foodpol.2020.101849.

Dunford, M. and Smith, A., 2000. Catching Up or Falling Behind? Economic Performance and Regional Trajectories in the New Europe. *Economic Geography*, 76(2), 169–95. https://doi.org/10.1111/j.1944-8287.2000.tb00139.x.

Dzenovska, D., 2018. *School of Europeanness: Tolerance and Other Lessons in Political Liberalism in Latvia*. Cornell University Press, Ithaca, NY.

Edelman, M., 1999. *Peasants Against Globalization: Rural Social Movements in Costa Rica*. Stanford University Press, Stanford, CA.

Engel-Di Mauro, S., 2009. Seeing the Local in the Global: Political Ecologies, World-systems, and the Question of Scale. *Geoforum*, 40(1), 116–25. https://doi.org/10.1016/j.geoforum.2008.09.004.

—— (Ed.), 2006. *The European's Burden: Global Imperialism in EU Expansion*. Peter Lang, New York.

Federici, S., 2012. *Revolution at Point Zero: Housework, Reproduction, and Feminist Struggle*. PM Press, Oakland, CA.

Fendrychová, L., and Jehlička, P., 2018. Revealing the Hidden Geography of Alternative Food Networks: The Traveling Concept of Farmers' Markets. *Geoforum*, 95, 1–10. https://doi.org/10.1016/j.geoforum.2018.06.012.

Fischler, F., 2000. A New CAP for a New Century. http://europa.eu/rapid/press-release_SPEECH-00-75_en.htm?locale=en.

Friedmann, H., 1978. World Market, State, and Family Farm: Social Bases of Household Production in the Era of Wage Labor. *Comparative Studies in Society and History*, 20(4), 545–86. https://doi.org/10.1017/S001041750001255X.

Galt, R. E., 2013a. Placing Food Systems in First World Political Ecology: A Review and Research Agenda. *Geography Compass*, 7 (9), 637–58. https://doi.org/10.1111/gec3.12070.

——, 2013b. The Moral Economy Is a Double-Edged Sword: Explaining Farmers' Earnings and Self-Exploitation in Community-Supported Agriculture. *Economic Geography*, 89(4), 341–65. https://doi.org/10.1111/ecge.12015.

Garratt, Elisabeth, 2020. Food Insecurity in Europe: Who is at Risk, and How Successful are Social Benefits in Protecting Against Food Insecurity? *Journal of Social Policy*, 49(4), 785–809. https://doi.org/10.1017/s0047279419000746.

Gibson-Graham, J. K., 1996. *The End of Capitalism (As We Knew It): A Feminist Critique of Political Economy*. University of Minnesota Press, Minneapolis, MN.

Goodman, D., and DuPuis, E. M., 2002. Knowing Food and Growing Food: Beyond the Production — Consumption Debate in the Sociology of Agriculture. *Sociologia Ruralis*, 42(1), 5–22. https://doi.org/10.1111/1467-9523.00199.

Goodman, D., DuPuis, E. M., and Goodman, M. K., 2012. *Alternative Food Networks: Knowledge, Practice, and Politics*. Routledge, New York.

Goodman, D., Fernando, S., and Wilkinson, J., 1987. *From Farming to Biotechnology: A Theory of Agro-Industrial Development*. Blackwell, New York.

Goszczyński, W., Śpiewak, R., Bilewicz, A., and Wróblewski, M., 2019. Between Imitation and Embeddedness: Three Types of Polish Alternative Food Networks. *Sustainability*, 11(24), 7059–78. https://doi.org/10.3390/su11247059.

Grivins, M., and Tisenkopfs, T., 2015. A Discursive Analysis of Oppositional Interpretations of the Agro-Food System: A Case Study of Latvia. *Journal of Rural Studies*, 39, 111–21. https://doi.org/10.1016/j.jrurstud.2015.03.012.

Guthman, J., 2004. Back to the Land: The Paradox of Organic Food Standards. *Environment and Planning A*, 36, 511–28. https://doi.org/10.1068/a36104.

Hall, D., 2013. Primitive Accumulation, Accumulation by Dispossession and the Global Land Grab. *Third World Quarterly*, 34(9), 1582–604. https://doi.org/10.1080/01436597.2013.843854.

Hann, C. M., (Ed.), 2003. The Postsocialist Agrarian Question: Property Relations and the Rural Condition. Lit Verlag Münster, Münster.

Hudson, R., 2003. European Integration and New Forms of Uneven Development: But Not the End of Territorially Distinctive Capitalisms in Europe. *European Urban and Regional Studies*, 10(1), 49–67. https://doi.org/10.1177/a032539.

Humphrey, C., and Mandel, R. (Eds), 2002. *Markets and Moralities: Ethnographies of Postsocialism*. Berg Publishers, London.

Jehlička, P., 2021. Eastern Europe and the Geography of Knowledge Production: The Case of the Invisible Gardener. *Progress in Human Geography*. https://doi.org/10.1177/0309132520987305.

Jehlička, P., Daněk, P., and Vávra, J., 2019. Rethinking Resilience: Home Gardening, Food Sharing and Everyday Resistance. *Canadian Journal of Development Studies*, 40(4), 511–27. https://doi.org/10.1080/02255189.2018.1498325.

Jehlička, P., Grīviņš, M., Visser, O., and Balázs, B., 2020. Thinking Food Like an East European: a Critical Reflection on the Framing of Food Systems. *Journal of Rural Studies*, 76, 286–95. https://doi.org/10.1016/j.jrurstud.2020.04.015.

Jung, Y., Klein, J. A., and Caldwell, M. L. (Eds.), 2014. *Ethical Eating in the Postsocialist and Socialist World*. University of California Press, Berkeley, CA.

Kautsky, K., 1988. *The Agrarian Question*. Zwan Press, London.

Kay, C., 1981. Political Economy, Class Alliances and Agrarian Change in Chile. *The Journal of Peasant Studies*, 8(4), 485–513. https://doi.org/10.1080/03066158108438148.

Katz, C., 2001. Vagabond Capitalism and the Necessity of Social Reproduction. *Antipode*, 33(4), 709–28. https://doi.org/10.1111/1467-8330.00207.

Kim, D. J., 2012. Taking Better Care of the Fields: Knowledge Politics of Sugar Beet, Soil, and Agriculture after Socialism in Western Poland. University of Michigan (Unpublished doctoral thesis).

——, 2011. Clutching the Ladder of Development: European Sugar Reform in Poland. *Anthropological Journal of European Cultures*, 20(1), 29–47. https://doi.org/10.3167/ajec.2011.200103.

Klumbytė, N., 2011. Europe and Its Fragments: Europeanization, Nationalisms, and the Geopolitics of Provinciality in Lithuania. *Slavic Review*, 70(4), 844–72. https://doi.org/10.5612/slavicreview.70.4.0844.

Kovács, E. K., 2015. Surveillance and State-Making Through EU Agricultural Policy in Hungary. *Geoforum*, 64, 168–81. https://doi.org/10.1016/j.geoforum.2015.06.020.

Kuus, M., 2004. Europe's Eastern Expansion and the Reinscription of Otherness in East-Central Europe. *Progress in Human Geography*, 28(4), 472–89. https://doi.org/10.1191/0309132504ph498oa.

Le Velly, R., and Dufeu, I., 2016. Alternative Food Networks as 'Market Agencements': Exploring Their Multiple Hybridities. *Journal of Rural Studies*, 43, 173–82. https://doi.org/10.1016/j.jrurstud.2015.11.015.

Lenin, V. I., 2004 [1899]. *The Development of Capitalism in Russia*. University Press of the Pacific, Honolulu.

Leonard, P., and Kaneff, D. (Eds.), 2002. *Post-Socialist Peasant? Rural and Urban Constructions of Identity in Eastern Europe, East Asia and the Former Soviet Union*. Palgrave Macmillan, New York.

Levin, R., and Neocosmos, M., 1989. The Agrarian Question and Class Contradictions in South Africa: Some Theoretical Considerations. *The Journal of Peasant Studies*, 16(2), 230–59. https://doi.org/10.1080/03066158908438391.

Łuczaj, L., Pieroni, A., Tardío, J., Pardo-de-Santayana, M., Sõukand, R., Svanberg, I., and Kalle, R., 2012. Wild food plant use in 21st century Europe, the disappearance of old traditions and the search for new cuisines involving wild edibles. *Acta Societatis Botanicorum Poloniae*, 81(4), 359–70. https://doi.org/10.5586/asbp.2012.031.

Mann, S. A., 1990. *Agrarian Capitalism in Theory and Practice*. The University of North Carolina Press, Chapel Hill, NC.

Massey, D., 2007. *World City*. John Wiley & Sons, Malden, MA.

——, 2005. *For Space*. SAGE Publications Ltd, London.

McMichael, P., 2009. A Food Regime Genealogy. *The Journal of Peasant Studies*, 36(1), 139–69. https://doi.org/10.1080/03066150902820354.

Mincyte, D., 2012. How Milk Does the World Good: Vernacular Sustainability and Alternative Food Systems in Post-Socialist Europe. *Agriculture and Human Values*, 29(1), 41–52. https://doi.org/10.1007/s10460-011-9328-8.

——, 2011. Subsistence and Sustainability in Post-Industrial Europe: The Politics of Small-Scale Farming in Europeanising Lithuania. *Sociologia Ruralis*, 51(2), 101–18. https://doi.org/10.1111/j.1467-9523.2011.00530.x.

Mincytė, D., Bartkienė, A., and Bikauskaitė, R., 2020. Diverging temporalities of care work on urban farms: Negotiating history, responsibility, and productivity in Lithuania. *Geoforum*, 115, 44–53. https://doi.org/10.1016/j.geoforum.2020.06.006.

Mroczkowska, J., 2019. Pork politics: The Scales of Home-made Food in Eastern Poland. *Appetite*, 140, 223–30. https://doi.org/10.1016/j.appet.2019.04.022.

Murdoch, J., Marsden, T., and Banks, J., 2000. Quality, Nature, and Embeddedness: Some Theoretical Considerations in the Context of the Food Sector. *Economic*

Geography, 76(2), 107–25. https://doi.org/10.1111/j.1944-8287.2000.tb00136.x.

Otero, G., 2018. *The Neoliberal Diet: Healthy Profits, Unhealthy People*. University of Texas Press, Austin, TX.

Pieroni, A., and Sõukand, R., 2018. Forest as stronghold of local ecological practice: currently used wild food plants in Polesia, Northern Ukraine. *Economic Botany*, 72(3), 311–31. https://doi.org/10.1007/s12231-018-9425-3.

Ponte, S., 2016. Convention Theory in the Anglophone Agro-Food Literature: Past, Present and Future. *Journal of Rural Studies*, 44, 12–23. https://doi.org/10.1016/j.jrurstud.2015.12.019.

Postone, M., 1993. *Time, Labor, and Social Domination: A Reinterpretation of Marx's Critical Theory*. Cambridge University Press, Cambridge.

Pungas, L., 2019. Food Self-Provisioning as an Answer to the Metabolic Rift: The Case of 'Dacha Resilience' in Estonia. *Journal of Rural Studies*, 68, 75–86. https://doi.org/10.1016/j.jrurstud.2019.02.010.

Ramamurthy, P., 2000. The Cotton Commodity Chain, Women, Work and Agency in India and Japan: The Case for Feminist Agro-Food Systems Research. *World Development*, 28(3), 551–78. https://doi.org/10.1016/S0305-750X(99)00137-0.

Razavi, S., 2009. Engendering the Political Economy of Agrarian Change. *The Journal of Peasant Studies*, 36(1), 197–226. https://doi.org/10.1080/03066150902820412.

Schwartz, K., 2007. 'The Occupation of Beauty': Imagining Nature and Nation in Latvia. *East European Politics & Societies*, 21(2), 259–93. https://doi.org/10.1177/0888325407299781.

——, 2005. Wild Horses in a 'European Wilderness': Imagining Sustainable Development in the Post-Communist Countryside. *Cultural Geographies*, 12(3), 292–320. https://doi.org/10.1191/1474474005eu331oa.

Seeth, H. T., Chachnov, S., Surinov, A., and Von Braun, J., 1998. Russian Poverty: Muddling through Economic Transition with Garden Plots. *World Development*, 26(9), 1611–24. https://doi.org/10.1016/S0305-750X(98)00083-7.

Smeds, J., 2015. Growing through Connections — A Multi-Case Study of Two Alternative Food Networks in Cluj-Napoca, Romania. *Future of Food: Journal on Food, Agriculture and Society*, 2(2), 48–61.

Smith, A., 2002. Imagining Geographies of the 'New Europe': Geo-Economic Power and the New European Architecture of Integration. *Political Geography, Special Issue Dedicated to Saul B. Cohen*, 21(5), 647–70. https://doi.org/10.1016/S0962-6298(02)00011-2.

Smith, A., and Stenning, A., 2006. Beyond Household Economies: Articulations and Spaces of Economic Practice in Postsocialism. *Progress in Human Geography*, 30(2), 190–213. https://doi.org/10.1191/0309132506ph601oa.

Smith, A., Stenning, A., Rochovská, A., and Świątek, D., 2008. The Emergence of a Working Poor: Labour Markets, Neoliberalisation and Diverse Economies in Post-Socialist Cities. *Antipode*, 40(2), 283–311. https://doi.org/10.1111/j.1467-8330.2008.00592.x.

Smith, J., and Jehlička, P., 2013. Quiet Sustainability: Fertile Lessons from Europe's Productive Gardeners. *Journal of Rural Studies*, 32, 148–57. https://doi.org/10.1016/J.JRURSTUD.2013.05.002.

Sõukand, R., Stryamets, N., Fontefrancesco, M. F., and Pieroni, A., 2020. The importance of tolerating interstices: Babushka markets in Ukraine and Eastern Europe and their role in maintaining local food knowledge and diversity. *Heliyon*, 6(1), e03222. https://doi.org/10.1016/j.heliyon.2020.e03222.

Spilková, J., Fendrychová, L., and Syrovátková, M., 2013. Farmers' Markets in Prague: A New Challenge within the Urban Shoppingscape. *Agriculture and Human Values*, 30(2), 179–91. https://doi.org/10.1007/s10460-012-9395-5.

Spilková, J., and Perlín, R., 2013. Farmers' Markets in Czechia: Risks and Possibilities. *Journal of Rural Studies*, 32, 220–29. https://doi.org/10.1016/j.jrurstud.2013.07.001.

Spilková, J., and Vágner, J., 2018. Food Gardens as Important Elements of Urban Agriculture: Spatio-Developmental Trends and Future Prospects for Urban Gardening in Czechia. *Norsk Geografisk Tidsskrift — Norwegian Journal of Geography*, 72(1), 1–12. https://doi.org/10.1080/00291951.2017.1404489.

Staddon, C., 2009. Towards a Critical Political Ecology of Human–Forest Interactions: Collecting Herbs and Mushrooms in a Bulgarian Locality. *Transactions of the Institute of British Geographers*, 34(2), 161–76. https://doi.org/10.1111/j.1475-5661.2009.00339.x.

Stuckler, D., and Basu, S., 2013. *The Body Economic: Why Austerity Kills*. Basic Books, New York.

Syrovátková, M., 2016. The Adoption of a Local Food Concept in Post-Communist Context: Farm Shops in Czechia. *Norsk Geografisk Tidsskrift — Norwegian Journal of Geography*, 70(1), 24–40. https://doi.org/10.1080/00291951.2015.1125942.

Syrovátková, M., Hrabák, J., and Spilková, J., 2015. Farmers' Markets' Locavore Challenge: The Potential of Local Food Production for Newly Emerged Farmers' Markets in Czechia. *Renewable Agriculture and Food Systems*, 30(4), 305–17. https://doi.org/10.1017/S1742170514000064.

Turner, M. D., 2017. Political Ecology III: The Commons and Commoning. *Progress in Human Geography*, 41(6), 795–802. https://doi.org/10.1177/0309132516664433.

Verdery, K., 2003. *The Vanishing Hectare: Property and Value in Postsocialist Transylvania*. Cornell University Press, Ithaca, NY.

Vergopoulos, K., 1978. Capitalism and Peasant Productivity. *The Journal of Peasant Studies*, 5 (4), 446–65. https://doi.org/10.1080/03066157808438057.

Wallace, R., 2020. *Dead Epidemiologists: On the Origins of COVID-19*. Monthly Review Press, New York.

Wallace, R. G, Kock, R., Bergmann, L., Gilbert, M., Hogerwerf, L., Pittiglio, C., Mattioli, R., and Wallace, R., 2016. Did Neoliberalizing West African Forests Produce a New Niche for Ebola? *International Journal of Health Services*, 46(1), 149–65. https://doi.org/10.1177/0020731415611644.

Watts, M., and Goodman, D. (Eds.), 1997. *Globalising Food: Agrarian Questions and Global Restructuring*. Routledge, London.

Whatmore, S., and Thorne, L., 1997. Nourishing Networks: Alternative Geographies of Food, in: Goodman, D., and Watts, M. (Eds.), *Globalising Food: Agrarian Questions and Global Restructuring*. Routledge, London.

Wills, B., and Arundel, A., 2017. Internet-Enabled Access to Alternative Food Networks: A Comparison of Online and Offline Food Shoppers and Their Differing Interpretations of Quality. *Agriculture and Human Values*, 34(3), 701–12. https://doi.org/10.1007/s10460-017-9771-n.

Wolff, L., 1994. *Inventing Eastern Europe: The Map of Civilization on the Mind of the Enlightenment*. Stanford University Press, Stanford, CA.

Woolfson, C., and Sommers, J., 2016. Austerity and the Demise of Social Europe: The Baltic Model versus the European Social Model. *Globalizations*, 13(1), 78–93. https://doi.org/10.1080/14747731.2015.1052623.

Yotova, M., 2018. The 'Goodness' of Homemade Yogurt: Self-Provisioning as Sustainable Food Practices in Post-Socialist Bulgaria. *Local Environment*, 23(11), 1063–74. https://doi.org/10.1080/13549839.2017.1420048.

Zbierski-Salameh, S., 1999. Polish Peasants in the 'Valley of Transition': Responses to Postsocialist Reforms, in: Burawoy, M., and Verdery, K. (Eds.), *Uncertain Transition: Ethnographies of Change in the Postsocialist World*. Rowman & Littlefield, Oxford.

Zhang, Q. F., 2015. Class Differentiation in Rural China: Dynamics of Accumulation, Commodification and State Intervention. *Journal of Agrarian Change*, 15(3), 338–65. https://doi.org/10.1111/joac.12120.

11. What Is Not Known about Rural Development? Village Experiences from Serbia

Jovana Dikovic

I centre this discussion around extended ethnographic research in three villages in Vojvodina Province in Serbia, where I studied understandings and practical effects of state-led and endogenous rural development. Intellectual conceptualisations of rural development and its realisation by state and policy makers have ceased to mean only the improvement of economic and social standing of land users (or targeted groups or individuals), but also imply rather complex planning, goals (Scoones, 2009) or ideology (Sachs, 2009) whose effects tie together economic, social and environmental improvements to all categories of people in rural areas. In contrast, endogenous rural development draws from cohesive community structures and represents both the source of social and individual identity and social obligations within and toward communities. While state- and donor-led rural development projects often fail to achieve their encompassing goals (see Blackburn and Holland, 1998; Mosse, 2001; Cooke and Kothari, 2001; Hobart, 1993; Chambers, 1983; Higgot, 1983), this chapter will unpack evidence from my fieldsites that suggests that endogenous rural development driven by local values, in combination with favourable market incentives, may achieve wider effects.

These effects develop for at least two reasons. First, there is a structural difference between state-led and endogenous rural development. Agricultural policies and measures conducted by the Serbian Ministry of Agriculture from 2000 until today, are rudimentary, predominantly

 https://doi.org/10.11647/OBP.0244.11

focused on the intensification of agricultural production, and also exclusive, targeting only registered agricultural producers who actively practice agriculture. Endogenous development, on the other hand, spreads horizontally and is more inclusive toward diverse categories of people and individual approaches to agriculture. Second, the ideology of agricultural policy-makers often does not comply with the ideology of producers, as evidenced in numerous agri-environmental and conservation schemes carried out across Europe (Burton et al., 2008; Medina et al., 2017). Within endogenous development, sharing local values and worldviews is in most cases a necessary precondition for an internal realisation of functional household and village organisation. Some local values in Serbian villages have proven to be resistant to challenges and change, particularly during socialism, and thus remain as cultural patterns that may be regarded as important factors that characterise some rural communities. In this paper, I provide insights into how localised understandings of emic concepts such as 'hard work' and 'dignity' are maintained as the main drivers of rural development.

Emphasis on particular values that are valorised by the community emerged from the analysis of qualitative interviews. Values around hard work and dignity for the majority who live in the village and work in agriculture are associated with keeping farms and households operative, productive and tidy. Similar to farmers in Vojvodina Province, among farmers in the UK "land becomes essential for the farm family to construct a 'farmer' identity, i.e. in a symbolic sense it becomes an integral part of the farmer" (Burton, 2004: 207). The missed opportunity to acquire land, or to maintain and enlarge existing capital is commonly associated with failure. Land enables the functioning of the established system for the display of virtuous behaviours through the role and widespread respect and esteem held for the character of the 'good farmer', where ideas around being productive and committed come from (Burton, 2004; Silvasti, 2003). Hard work and dignity are therefore local valorised values. While 'hard work' is both a symbolic and productive expression of a farmer's commitment to agriculture (Burton, 2004; Emery, 2014), dignity is held to be its result. Dignity, in other words, emerges from "honest sweat", invested effort and time in tillage of owned or leased land (James, 1899: 262, cited in Burton, 2004: 197), and can also be extended to a sense of personal autonomy and liberty common to farmers all over the world (Stock et al., 2014).

Yet the importance of emic ideas about how communities define well-being, and how local forms of 'thriving' may be achieved get neglected within bureaucratic 'improvement' of the rural condition because of mistaken premises around how change occurs. Policy-makers' ideas about change in rural areas are first formalised through project plans, laws, and financial institutions (Apthorpe, 1997). Rural development projects are, for this reason, predominantly influenced by a neo-institutional theoretical paradigm, because "institutions (most commonly conceptualised as organisations) are highly attractive to theorists, development policy-makers and practitioners as they help to render legible 'community' and codify the translation of individual into collective endeavours in a form that is visible, analysable and amenable to intervention and influence" (Cleaver, 2001: 40). Thus, it is usually thought that further planning, bureaucratic adjustments and new regulations may enable targeted areas to better 'thrive'. An emphasis on planning does not tell us *how* rural development in fact occurs, on the ground. This is problematic as in most cases, rural development does not take place in expected, nor in institutionally-controlled, ways.

By not acknowledging local ideas and understandings of development, life beyond state planning, acts and regulations remains unknown, and the values that underpin the society in question are neglected (Pandian, 2009). "Neglect of the subjective dimension of value makes it difficult to understand cultural reproduction, or, for that matter, change" (Robbins and Sommerschuh, 2016: 6). Explanation of the process of change on the basis of virtues, ethics and rhetoric is a sophisticated endeavour that requires a very committed, grounded methodology (see for example McCloskey, 2006, 2016a). Here I restrict my attention to analyses of how 'hard work' and 'dignity', as they emerge from committed agricultural work and the acquisition or maintenance of land ownership, may spur local development.

As already indicated, in some parts of the world land and land ownership has played a structural role in identity and personhood formation in rural settings (Burton, 2004; Macfarlane, 1979). The same observation applies to my fieldsites where the importance of land ownership represents a materialistic and symbolic framework for the expression and practice of extant village values.

While it could easily be argued that those without land and capital are left out of definitions of local economic development, I make the case here that local development is more inclusive in an ideological and practical sense than that driven by the Serbian Ministry of Agriculture, because it is based on the principal application of norms that hold local society together, unlike the selective bureaucratic application of agricultural schemas. Likewise, thriving 'from scratch' in the local context is possible and common, unlike the existing Serbian state agricultural schemes driven by administrative mechanisms that predominantly exclude those who do not use or possess land.

Therefore, in this chapter, I investigate how local visions of development are constituted, and dissect them to examine their elements of 'hard work' and 'dignity' in relation to labour and land ownership. Such a perspective provides an important understanding of the Serbian context as it lies on the cusp of introducing widespread agricultural support, as part of enormous transformations that will inevitably change the social climate in these communities and the sector in the years to come. After the introduction to the research conducted, I briefly explain the state of the agricultural sector in Serbia after 2000, with particular attention to state-led rural development programmes. In the next section, I provide insights on endogenous rural development and explain why hard work and dignity matter in the context of the villages studied, and then I shift to three narratives that illustrate how these values have determined individual achievements. In the final section, I provide an explanation of endogenous rural development and point to why it is structurally different from formal conceptualisations and realisations of rural development programs.

Development through Ethnographic Research

Agricultural producers from the villages of Gaj, Beli Breg and Malo Bavanište make up the core of my investigation. The majority of land users cultivate between 5 and 20 ha, while a somewhat smaller though significant group cultivates 30–60 ha. Only a few informants cultivate more than 70 ha and they represent the wealthiest producers in the villages. During my extensive research, from February to September 2013, with several visits thereafter in 2014, 2015, and 2017, I conducted qualitative semi-structured interviews with more than ninety people of different backgrounds, and observed and participated in daily life within these villages.

My interviews focused on local understandings of development and values and how these related to state agricultural policies (subsidies, leasing of state land, EU pre-accession instruments for rural development) and their effects on the producers and their livelihoods. On different occasions, people often referred to hard work, dignity, self-realisation, or problems in fulfilling these local values (alcoholism, divorce, problematic personalities, laziness, lack of commitment, etc.). These were also seen as drivers and as evaluative criteria for the assessment of others' work or commitment.

Based on these qualitative interviews and extensive participant observation, I selected sixteen interviews that I evaluated to be the most developed articulations of values in relation to land and farming. Even though the practical influence of these selected cases is difficult to estimate, they nevertheless mirror an identified local trend and worldview of dwellers for whom agriculture is a prime source of income. In this chapter, I examine three individual stories out of this narrow group that share some paradigmatic characteristics of observed endogenous rural development, and reflect a grounded variety of individual approaches to agriculture and motivations that stem from different ethnic, social and economic backgrounds. These stories also reveal the potential of social imitation of locally held values, which when attained, apart from improving individuals' economic standing, signals a commitment to agriculture and a continuation of the village worldview.

The State of Agriculture and Rural Development

Throughout the twentieth century, Serbia experienced two significantly different agricultural reforms that initiated colossal changes in property structure and agricultural organisation. By the first agrarian reform (1919–1941), the Kingdom of Yugoslavia eliminated any remains of the feudal ownership structure by inaugurating small private landholdings and facilitated capitalist production relationships in agriculture. By the second agrarian reform (1945–1953), in socialist Yugoslavia, unsuccessful attempts at collectivisation and very poor functionality of cooperatives meant that private landholdings were not prohibited but limited to 10 ha, while the cornerstone of agricultural organisation became complex, state-run, agri-industrial systems. Such regulations enabled continuity of private landholdings throughout the socialist period including a variety of informal strategies for acquiring additional land (Dikovic, 2015).

In 1991, the Law on Restitution initiated a new phase in the rehabilitation of private property rights that established agriculture on a market economy basis. By then, three forms of ownership and production had existed: private, state, and social. But since then, apart from continual development of private property relations, significant state ownership in agriculture has continued to be a source of strong debates, because of acreages that have not yet been restituted, and postponed privatisation in the name of political goals that has led to monopolies, clientelism and rent-seeking (Maksimović-Sekulić et al., 2018; Pejanović et al., 2017).

In 2000 with the first democratic government, the Serbian Ministry of Agriculture initiated steps toward EU integration, and soon completed the registration of agricultural households to map the active agricultural and rural population. As other accounts from eastern European member states show, the ministry imported a rather homogeneous body of agricultural and later rural development policies, laws and trade agreements, with the aim of professionalising agriculture based on western European intensive agricultural models (Diković, 2014). The ministry was driven by pragmatic interests toward enhancing production, because of many factors that were seen as unfavourable, rather than by the improvement of environmental and social conditions in rural areas. For this reason, 80–88% of the total agricultural budget is devoted to direct subsidy programs to intensify production, without reference to environmental protection or conservation (Ćurković, 2013).

Subsidies were introduced in 2006, and peaked during 2008–2013 when farmers who cultivated up to 99 ha had the right to claim 120 EUR/ha, thereby providing an advantage to those farmers who already had better economic standing. As of 2016, the Ministry of Agriculture reduced subsidies to 35 EUR/ha and limited them to those who cultivate only 20 ha, in response to "failed policies that did not give the expected results".[1] Thus, subsidies today represent a form of state-social

1 See: http://rs.n1info.com/Biznis/a131806/Subvencije-poljoprivredi-smanjene-zbog-zloupotreba.html and https://www.rts.rs/page/stories/sr/story/13/ekonomija/1493441/manje-subvencije-za-ratare-ko-placa-ceh.html. Translation my own.

assistance in agriculture rather than an incentive for production, and in their current form do not 'motivate' work. For small producers, subsidies ensure survival in the business and keep their heads above water. Unlike other EU countries where subsidies make up more than half the income of farmers whose behaviour and decisions largely depend on state support (Sutherland, 2010; Medina et al., 2015), in Serbia in 2012 subsidies made up 8% of the gross income of middle-sized agricultural producers. Today this figure is likely to be even less due to further reductions in subsidy.[2]

Further, the Instrument for Pre-accession Assistance (IPA), and the Instrument for Pre-accession Assistance for Rural Development (IPARD) are aimed at all types of agricultural households, but the application rates are very low, because they require pre-investment and developed business plans from farmers, which many producers are not willing to make (Milovanović, 2016). Such a situation reveals in the first place that Serbian agriculture has not yet developed the culture and knowledge associated with the EU's Common Agricultural Policy (CAP), from either state planners or end-users (Papić and Bogdanov, 2015). Pre-accession funds in Serbia are based on competitive platforms that favour better-off and advanced agricultural producers, unlike other EU programmes that have been initiating new agricultural and environmental schemes that are also competitive but target different types of actors in agribusiness.

Thus, the 'inbetweenness' of Serbian agriculture—with its policies that neither meet farmers' needs nor satisfy EU standards, with strong state control over its capital in agriculture on the one hand, and market-oriented low subsidies on the other, with the economic stagnation of one part of agricultural households alongside an ongoing, small, but important agricultural revolution at other local levels (SEEDEV, 2017)—creates conditions that have many positive and negative sides in the context of the rural development. Within such a state of 'inbetweenness', people transition, explore and take forward local norms surrounding agriculture. Conceptualisations of 'good' farmers, 'hard' work and dignity have a long, value-laden history in the region, and their modern manifestations are explored below.

2 See http://www.agroservis.rs/poslovanje-poljoprivrednog-gazdinstva-2012-godine.

Insights on Endogenous Rural Development

I structurally altered my understanding of the role of the state in shaping rural, economic, and social environments after one essential statement from an informant:

> Land never stays uncultivated, no matter the political and economic conditions. During the worst period of Serbian agriculture and economic sanctions in the 1990s people were selling a bull or a barrel of oil to cultivate the land, and, surprisingly, not a single hectare remained uncultivated at that time (Jovan, July 2014).

This observation contains the powerful claim that land-people relationships transcended the political-economic system, and determine life in the village more than any other wider institutional transformation or planned change. The following example illustrates it further.

> Marko: During the 1990s I paid 400 DM for a barrel of oil for cultivation of the land.
>
> J. D.: Oh! Isn't that too expensive for one barrel? I mean, why would you pay so much for a single barrel of oil?
>
> Marko: Well, a person who has grown up here cannot leave the land uncultivated. You watch it every day. You can't let it go. It's in the genes. Either you love it or you don't. And yes, it cost me like Greece [a common Serbian expression when someone runs into debt], but, well, we are used to living like this. I was working at a loss back then, but at least my children weren't hungry, and we didn't lack anything in the household. (July 2017)

When, during an informal talk, I asked one farmer how it felt to be in the fields, he simply said, "I feel great!" (Jaroslav, March 2013). His prompt and instinctive answer insinuated that land to him represented a symbiotic relationship between ownership, obligation, emotional attachment and professional satisfaction. Arguably, these cannot be structurally shaken, even by external factors. Farmers' dedication to land and agriculture was also recognised by a local doctor:

> When patients go home from hospital, they first go to visit the fields, to check the quality of corn, wheat. It is their life, and more. It is their very love for the land. (Marija, June 2013)

It is well-documented that a crucial aspect of private property lies in control, not only over one's own resources, but also over one's own life, even though it may be a liability and a burden at the same time (Sikor, 2006).

Fig. 1. Threshing in Gaj. Photograph by the author (2014).

The social burden of land and the maintenance of the farm household are both a constraining and rewarding experience. The failure to do so might affect the reputation of a person (and their family). Meeting locally-set expectations, in combination with individual aspirations for betterment, may also be rewarding to an individual (and their family), and to their social and symbolic standing. To earn a good reputation, one cannot go below an implied local standard which requires synergic interaction of the deadlines imposed by nature, followed by hard work and appropriate treatment of the land that meets the productive, aesthetic and sentimental expectations of the producer and community. The standard of meeting such local norms provides a person with a sense of dignity and accomplishment and serves as a positive example motivating others to do the same. This explains why the land never stays uncultivated, even when it makes an absolute loss, and points to the idea of endogenous rural development which may proceed counterintuitively to the economic and political circumstances. The following protagonists emphasise the practical importance of local social climate and understanding of what it takes to be subjectively and socially regarded as a virtuous and productive farmer and co-villager.

The Local Vision of Development: Three Stories

When Maruška (aged fifty-four) and her husband married they did not have anything. They lived in one improvised room without a bathroom under the barn together with their two small sons. In the beginning, they worked as day labourers mostly in private households and the state agricultural farm in the village. From the money they saved, they bought a small plot of land and built a house. Then, they started a farm with fifteen pigs. Soon the farm started to grow and at the end of the year, they had sixty pigs. Since the farm was expanding, they moved it to the summer ranch near the village and lived there for five years together with their sons. But their sons could not accept their family business, which they found embarrassing, and asked them to shut down the farm because other people in the village looked down on them and called them "piggers" (*svinjari*).

Maruška and her husband eventually sold off the pigs and from that money, they bought 3 ha of land, two tractors and have since committed exclusively to agriculture. During the last thirty years, they acquired all the necessary mechanisation, a total 10 ha of their own land, and, moreover, they have taken a further 7 ha on lease. In the meantime, they expanded their household activities with cows, sheep, and pigs. Maruška and her husband built two houses for their two sons and organised two generous send-off parties for them when they joined the army. When their sons got married they organised big wedding ceremonies. She thinks all this would not have been possible without the pig business: "We thrived thanks to pigs, hard work, and our own sweat."

Although Maruška, her husband, sons, and their families live in two separate houses, they work the land and run the household together, and share all income equally. She is very proud of what they have achieved over the last thirty years and compared their success to the old well-known landowners (*gazde*)[3] from the village who failed.

> *Gazde* for whom we had previously worked for wages have ruined themselves, while we have thrived. This is because they strictly stick to the old, traditional, way of doing things, and the mothers-in-law have the last word. In our household we do it differently, we have a democratic approach, and we agree upon who is doing what. (July, 2013).

Maruška feels herself to be an example of a person whose success and well-being came as a result of hard work, a desire for a better life and social recognition. The importance of the latter was subtly emphasised in her talk about big celebrations and enormous spending, which are local indicators of status that satisfy both the desire for showing off on the one hand, and a public confirmation of one's success on the other.

***

Since he was ten years old, Goran (aged forty-three) has worked in agriculture and has always seen himself in it. Goran attempted a small private business during the 1990s in order to diversify his sources of income and support the household. He bought a truck and did unreported transport services for ten years, but despite this, he never ceased to work the land.

By the 1990s Goran and his father had 2 ha of land and took an additional 15–20 ha on lease. From the 1990s to 2000, due to the political situation in the country, the price of land became cheaper, and Goran and his father seized this opportunity and started to buy 2–3 ha annually.

In 1995 Goran significantly advanced the household when he initiated informal cooperation with his neighbour and friend. The aim of their cooperation was to help each other out in different agricultural work and to combine machines. Their cooperation resulted in mutual benefit and prosperity and during this time Goran invested in expanding the

---

3 Most of my interlocutors have a common understanding of what it means to be a good *gazda*. A *gazda* is a person who is not neglectful and who puts jobs in the household and fields ahead of everything else; someone who is a good planner and whose household has all that is needed; whose household, fields and stalls are tidy; who is a good and generous host on celebrations; who respects local tradition and people in the community; who is honourable and content with the profession of farmer. Although there are several local synonyms, what distinguishes this from other terms, is that one cannot be a *gazda* without land.

stock of agricultural buildings. In 2005, Goran and his partner realised that they were now experienced and capable enough to run independent businesses and decided to end their partnership.

According to Goran, hard work, entrepreneurial initiatives, economic intuition, cooperation with his partner, as well as the small business that he ran for ten years, all helped him to invest in his production and accumulate land when it was cheap. He grew into one of the most successful agricultural producers in the village and today he has 60 ha of his own land and an additional 60 ha on lease. In 2013 he was given an award for being one of the most successful agricultural producers in the area.

Goran argues that most successful producers like him, who seized the opportunity and invested in their businesses, already stood out before the implementation of state programs such as subsidies. In his opinion, readiness to invest in production and to expand one's property are the key factors for development.

> Today the land is in the hands of those who want to work it properly, and many have thrived as *gazde*. Nowadays all producers compete to work better; much attention is given to proper care and treatment of the land. Some twenty people from the village have thrived and cultivated from 20 to 90 ha of their own land. (August, 2013)

The symbolic value of good harvest and well maintained fields, apart from their economic reward and valued aesthetic also demonstrate a farmer's commitments to soil and agriculture, which is positively evaluated and appreciated by his peers. "The symbolic value of the crop is thus in that it displays the farmer's commitment to agriculture as a way of life, to the soil and to the crop, and not in its display of the profitability of the farm" (Burton, 2004: 209). Likewise, the social value of competition spurs the enthusiasm for production, and in a synergetic way influences attitudes toward the intensification of production, thereby influencing local development trajectories.

***

Ivica (aged forty-three) and his family lived for a long time as tenants in relative poverty because of his father, an alcoholic who had a low reputation in the village. Despite being an excellent student in high school, Ivica was not allowed to go to university because his family had neither land nor money to support his ambitions. Ivica instead had to do various jobs. He worked as a day labourer, then in private manufacture of fruit juices, he drove a taxi, sold and smuggled cigarettes, and acted as a bingo caller. He sometimes worked twelve-hour days, and despite this harsh life he never despaired because his optimism and vision kept him strong and persistent.

Today Ivica works in a company in the nearby town. His hard work and commitment to the job have led him to the position of the chief of the department with a competitive salary. But the general situation in the company among employees makes Ivica very unhappy. Even though this position enabled him to travel and see the sea for the first time in his life, to refurbish his house, and to buy some land, he could not countenance staying in the company.

After nine years spent in the company, Ivica prepared to resign and commit completely to agriculture, and he started to buy agricultural machinery. During this period, he bought approximately 5 ha of land, expanded the garden and invested in poultry. He found his motivation in acquiring land and gaining the symbolic status of a respected householder. "Every time I bought a piece of land, I had to fast for a whole year to pay off the debt, but one motive has always kept me going—that a poor man should earn capital—because my parents did not have a gram of land" (February, 2013).

Even though Ivica cannot be compared to other, bigger producers, his case is particular because his dedication to agriculture allows him to build a respected name—where his father failed—and to erase the bad image that the village associates with his family. His religiosity also strengthens his social reputation, as was symbolically acknowledged when he was offered a place on the church council. His proven commitment to his faith and related good reputation turned out to be crucial factors on a few occasions when Ivica purchased land from people who did not want to sell it to other parties, but did want to sell it to him. They were even happy to wait for his payment because they trusted him and wanted to help him in fulfilling his ambitions.

Ivica's story represents the importance of 'classical' values that carry weight in rural Serbian society such as hard work, dedication to agriculture, and religiosity. These features were important in earlier times, and still represent key values in contemporary village life.

Explaining Endogenous Development Stories

The living conditions of the individuals portrayed above have significantly improved in recent decades. They have thrived economically by acquiring and enlarging their existing capital. Common to all of these personal histories is that their improvement dates back to before state-led agricultural support. According to these individuals, existing state measures, including subsidies, did stimulate their production, but they did not trigger their desire nor their achievements. Likewise, favourable economic conditions and good prices on produce have also boosted their professional satisfaction and motivation.

Fig. 2. Household in Gaj. Photograph by the author (2013).

These individuals' development was self-initiated, and motivated by the importance they placed on the pursuit and attainment of social distinction through locally recognised forms of 'hard work' and 'dignity', including building a good name or benevolence, competition, cooperation, and the desire for professional and personal autonomy made possible through land ownership. These values resemble those of neo-agrarianism, a social and political philosophy which places strong emphasis on private ownership that develops an owner's responsibilities as exercised through economic, social, environmental, and political spheres (Thompson, 2010; Fiskio, 2012). The norms that, thus, privilege and value ownership, integrity, labour, active work on the land, and attachment to a community, generate a cultivar of moral character (Berry, 1996). As in various other cases worldwide, the village and, consequently, the ethics of producers have derived directly from labour and agriculture.

> For centuries, the entire life of a peasant[4] went by in scheduled labour. His habits were formed according to this schedule. Every task, when it became due, was urgent. Every task and every deadline was vital. Not by the coercion of another man, but under the discipline imposed by nature and put on him directly. Such necessity always gives a high moral sense to the drudgery of the peasantry. Never is a peasant so ethical as in his work. The qualities thus gained by each generation are bequeathed to the next generation, who enrich the heritage. Rich funds of certain working qualities were thus accumulated and settled in the rural population. (Vukosavljević, 1983: 418, translation by the author)

Another comparable example comes from Emery (2014) who explores understandings of work ethics, specifically hard work, among British farmers who value 'hard work' even when their financial returns continue to fail. Work ethic, therefore, should not be considered only in a utilitarian sense, as a path toward the maximisation of production, profit and wellbeing, as it also influences identity formation. To that end, "meeting obligations, securing identity, status and structure, are as fundamental to livelihood as bread and shelter" (Wallman, 1979: 7). Burton also clearly points out that "while there are economic benefits to maintaining tidy agricultural landscapes, the social symbolic value of the tidy landscape is unlikely to be related solely to economic factors for the simple reason that while the tidy fields represent the output of agriculture, they do not indicate the level of inputs" (Burton, 2004: 209).

As much as hard work, dignity, accompanying elements (such as land ownership, autonomy, individual desires and motivations), and 'exemplary' people give rise to a particular type of valorised personal

---

4 In the original text, the author used the word *seljak*, which is a term for the peasant, still in wide use in rural Serbia.

Fig. 3. Family greenhouse. Photograph by the author (2017).

and family model, they equally influence the logics of local economic development. Thinkers of the utilitarian tradition (Bentham, [1781] 2000) would find many instrumental arguments to explain how development comes into being. These are partially right. The role of the local community in assessing, acknowledging, or disapproving of somebody's success might sometimes be decisive. Success may indeed be individual, but it is meaningful only in the community, which is why its maintenance implies certain liabilities on the part of those who are considered exemplary. Thus, spreading knowledge, information, and support to peers is not only a pure act of benevolence, or an ethical aspiration, but also an unspoken expectation put on those interested in maintaining their social standing in the village. Values, consequently, become instilled in people not only through the influence of collective representations of what is considered good and important in life (Robbins, 2015). Values become represented to and instilled in subjects through the influence of those considered to be "exemplary persons" (see Humphrey, 1997; Robbins and Sommerschuh, 2016). These persons, in fact, shape the community and impose their standards, unlike subsidies, IPA, IPARD and other state agricultural policies which in the local context seem irrelevant.

McCloskey's insights are valid here: "It won't suffice, as the World Bank nowadays recommends, to add institutions and stir" (2016b: 10). The idea and importance of 'thriving' in rural communities first occurs in people's mindsets. The value attributed to agriculture, assumed and related ethics on working the land and maintaining the household, signify that in these communities in Serbia, hard work (as well as the motivation to undertake it) is socially respected. Such a social climate spurs on some members in the community towards land, productivist orientations, but also towards liveability. It is true that Goran who cultivates 120 ha, and Ivica who cultivates 5 ha cannot be compared and will never be (financially) equal. Yet, what is common to both is their commonly held motivation and their emphasis on becoming a 'good' and 'dignified' householder, which accounts for their prosperity. The growth, therefore, cannot be measured only by quantifiable factors, but also by subjective dimensions of value and wellbeing.

Likewise, those who own between 5 and 10 ha do not consider themselves unsuccessful, nor poor. On the contrary: Maruška, as well as Ivica, see themselves as slowly thriving and living a dignified life. This is due to a combination of factors such as better prices for products, the enthusiasm for production, the desire for professional and personal autonomy obtained through land ownership, and the desire for a better life.

Conclusion

Most externally-introduced rural development projects are guided by noble and benevolent intentions. They however often fail to accomplish their goals. Planners are seldom interested in understanding whether and how their ideas of development and local values synchronise, and how, why and in which direction local populations internalise and modify extant values. Moreover, theoretical but also policy horizons become blurred as they cannot see the solution to the dilemma—"what, if not through planning?"

As shown in this account through examples from Serbian villagers, the role of hard work and dignity, exemplary persons, and village ethics drive some forms of economic activity and development, in particular around land and farming. The cases provided here do not just present the importance of individual thriving but also of the principle of subsidiarity that places modest responsibilities for local wellbeing at the lowest level of the community and through role models. Villagers 'develop' and undertake work with constant reference to neighbours and other villagers. In the village, nothing can compare to the intensity with which laziness is despised, and conversely the admiration for a hard-working householder. Similarly, neglected fields that urge for competent treatment by the farmer are considered as "an anathema to farmers' sense of their professional identity and expertise" (Burton, 2004: 208). Such local perceptions have created a self-regulating system, with elements of competition, judgement and shaming, in which neighbours and co-villagers look at each other, learn from each other, and compete with each other. The synergy of their activities creates the phenomenon of social imitation. There is an ethics of becoming 'better', not necessarily richer; by extension, there is attention given to how to improve land cultivation.

Farmers create their own progress, impose new standards, and change themselves accordingly. Their ethos, understanding of life, and their nerve seem apt for such achievements. Without the strong influence of state policies and subsidies on the daily decisions of farmers' businesses such a scenario of development is highly likely. If, in the course of joining the EU, the Serbian Ministry of Agriculture initiates the intense integration of Serbian agriculture with the Common Agricultural Policy, this will lead to an enormous influx of EU subsidies, where farmers' sense of autonomy and control might be limited or traded for subsidies (see Krasznai-Kovacs, 2019). Subsequent crowding-out of personal incentives, local motors of development and accumulated knowledge will become a justifiable concern along with the disappearance of the village social fabric.

If policy-makers acknowledge studies beyond their narrow plans and investigate local manifestations and forms of rural development, future interventions may be transformed in targeted areas. In other words, rural development may be undertaken with closer attention to local 'ways of doing things'. Moderating policy and regulations to rural conditions on-the-ground and recognising existing social climates and local visions of development can potentially offer better pragmatic ideas than the reinvention of the wheel and the application of homogenous solutions, as well as gaining the systematic momentum many governments hope for.

References

Apthorpe, R., 1997. Writing Development Policy and Policy Analyses Plain or Clear: On Language, Genre and Power, in: Shore, C., and Wright, S. (Eds.), *Anthropology of Policy: Critical Perspectives on Governance and Power*. Routledge, London and New York, pp. 34–45.

Bentham, J., (1781) 2000. *An Introduction to the Principles of Morals and Legislation*. Reprint, Batoche Books, Kitchener.

Berry, W., 1996. *The Unsettling of America: Culture & Agriculture*. Counterpoint, Berkeley.

Blackburn, J., and Holland, J., 1998. *Who Changes? Institutionalising Participation in Development*. Intermediate Technology Publications, London. https://doi.org/10.3362/9781780446417.

Burton, R. J. F., 2004. Seeing Through the 'Good Farmer's' Eyes: Towards Developing an Understanding of the Social Symbolic Value of 'Productivist' Behaviour. *Sociologia Ruralis*, 44 (2), 195–215. https://doi.org/10.1111/j.1467-9523.2004.00266.x.

Burton, R. J. F., Kuczera, C. and Schwarz, G., 2008. Exploring Farmers' Cultural Resistance to Voluntary Agri-environmental Schemes. *European Society for Rural Sociology*, 48 (1), 16–37. https://doi.org/10.1111/j.1467-9523.2008.00452.x.

Chambers, R., 1983. *Rural Development: Putting the Last First*. Longman, London. https://doi.org/10.4324/9781315835815.

Cleaver, F., 2001. Institutions, agency and the limitations of participatory approaches to development, in: Cooke, B., and Kothari, U. (Eds.), *Participation: The New Tyranny?* Zed Books, London, pp. 36–55.

Cook, B. and Kothari, U., 2001. *Participation: The New Tyranny*? Zed Books, London.

Ćurković, V., 2013. Finansijska podrška države proivodnji organske hrane—velika razvojna šansa Srbije. Univerzitet Singidunum, Belgrade (Unpublished doctoral thesis).

Dikovic, J., 2015. The Practices of Landownership in Vojvodina: The Case of Aradac, in: Siegrist, H., and Müller, D. (Eds.), *Property in East Central Europe: Notions, Institutions, and Practices of Landownership in the Twentieth Century*. Berghahn Books, New York, pp. 268–88.

——, 2014. Neither Peasant, Nor Farmer. Transformations of Agriculture in Serbia after 2000, in: Dorondel, S., and Serban, S. (Eds.), *At the Margins of History. The Agrarian Question in Southeast Europe.* Special issue of *Revista Martor* 19, Bucharest: Muzeul Național al Țăranului Român, pp. 149–62.

Emery, S. B., 2014. Hard work, productivity and the management of the farmed environment in anthropological perspective, in: Hamilton, L., Mitchell, L., and Mangan, A. (Eds.), *Contemporary Issues in Management*. Edward Elgar Publishing, Cheltenham, pp. 90–104.

Fiskio, J., 2012. Unsettling ecocriticism: rethinking agrarianism, place, and citizenship. *American Literature,* 84 (2), 301–25. https://doi.org/10.1215%2F00029831-1587359.

Higgott, R. A., 1983. *Political Development Theory: The Contemporary Debate*. Croom Helm, London.

Hobart, M., 1993. *An Anthropological Critique of Development: The Growth of Ignorance*. Routledge, London.

Holland, J., and Blackburn, J., 1998. *Whose Voice? Participatory Research and Policy Change*. Intermediate Technology Publications, London.

Humphrey, C., 1997. Exemplars and rules: aspects of the discourse of moralities in Mongolia, in: Howell, S. (Ed.), *The Ethnography of Moralities*. Routledge, New York, pp. 25–47.

Krasznai-Kovacs, E. 2019. Seeing subsidies like a farmer: emerging subsidy cultures in Hungary. *The Journal of Peasant Studies,* 48(2), 387–410. https://doi.org/10.1080/03066150.2019.1657842.

Macfarlane, A., 1979. *The Origins of English Individualism: The Family, Property and Social Transformation*. Cambridge University Press, Cambridge.

Maksimović-Sekulić, N., Živadinović, J., and Dimitrijević, L., 2018. Concerns about hamonization process of Serbian agricultural policy with EU standards. *Economics of Agriculture*, 65 (4), 1627–39. https://doi.org/10.5937/ekopolj1804627m.

McCloskey, D., 2016a. *Bourgeois Equality: How Ideas, Not Capital or Institutions, Enriched the World*. University of Chicago Press, Chicago, IL and London.

——, 2016b. Max U versus Humanomics: a critique of neo-institutionalism. *Journal of Institutional Economics,* 12(1), 1–27. https://doi.org/10.1017/s1744137415000053.

——, 2006. *The Bourgeois Virtues: Ethics for an Age of Commerce*. University of Chicago Press, Chicago, IL and London. http://dx.doi.org/10.7208/chicago/9780226556673.001.0001.

Medina, G., and Potter, C., 2017. The nature and developments of the Common Agricultural Policy: lessons for European integration from the UK perspective. *Journal of European Integration*, 39(4), 373–88. https://doi.org/10.1080/07036337.2017.1281263.

Medina, G., Potter, C., and Pokorny, B., 2015. Farm business pathways under agri-environmental policies: Lessons for policy design. *Estudos Sociedade e Agricultura*, 23(1), 5–30.

Milovanović, M., 2016. *Mogući efekti integracije u EU na poljoprivredu Srbije*. Univerzitet u Novom Sadu (Unpublished doctoral thesis).

Mosse, D., 2001. 'People's knowledge', participation and patronage: operations and representations in rural development, in: Cook, B., and Kothari, U. (Eds.), *Participation — The New Tyranny?* Zed Press, London, pp. 16–35.

Pandian, A., 2009. *Crooked Stalks: Cultivating Virtue in South India*. Duke University Press, Durham, NC. https://doi.org/10.1215/9780822391012.

Papić, R., and Bogdanov, N., 2015. Rural development policy: A perspective of local actors in Serbia. *Ekonomika Poljoprivrede*, 62(4), 1079–93. https://doi.org/10.5937/ekopolj1504079p.

Pejanović, R., Glavaš-Trbić, D., and Tomaš-Simin, M., 2017. Problems of agricultural and rural development in Serbia and necessity of new agricultural policy. *Economics of Agriculture*, 64 (4), 1619–33. https://doi.org/10.5937/ekopolj1704619p.

Robbins, J., 2015. Ritual, value, and example: on the perfection of cultural representations. *The Journal of the Royal Anthropological Institute*, 21(51), 18–29. https://doi.org/10.1111/1467-9655.12163.

Robbins, J., and Sommerschuh, J., 2016. Values, in: Stein, F., Lazar, S., Candea, M., Diemberger, H., Robbins, J., Sanchez, A., and Stasch, R. (Eds.), *The Cambridge Encyclopedia of Anthropology*. University of Cambridge, Cambridge. http://doi.org/10.29164/16values.

Sachs, I., 2009. Revisiting Development in the Twenty-First Century. *International Journal of Political Economy*, 38 (3), 5–21. https://doi.org/10.2753/ijp0891-1916380301.

Scoones, I., 2009. Livelihoods perspectives and rural development. *The Journal of Peasant Studies*, 36 (1), 171–96. https://doi.org/10.1080/03066150902820503.

Scott, J., 1977. *The Moral Economy of the Peasant. Rebellion and Subsistence in Southeast Asia*. Yale University Press, New Haven, CT.

SEEDEV, 2017. Konkurentnost poljoprivrede Srbije. Beograd. https://www.seedev.org/publikacije/Konkurentnost_poljoprivrede_Srbije/Konkurentnost_Srbije_Analiza.pdf.

Sikor, T., 2006. Land as Asset, Land as Liability: Property Politics in Rural Central and Eastern Europe, in: Benda-Beckmann, F. von, Benda-Beckmann, K. von, and Wiber, G. M. (Eds.), *Changing Properties of Property*. Berghahn Books, New York and Oxford, pp. 106–26.

Silvasti, T., 2003. The cultural model of "the good farmer" and the environmental question in Finland. *Agriculture and Human Values*, 20 (20), 143–50. https://doi.org/10.1023/a:1024021811419.

Stock, P. V., and Forney, J., 2014. Farmer autonomy and the farming self. *Journal of Rural Studies*, 36, 160–71. https://doi.org/10.1016/j.jrurstud.2014.07.004.

Sutherland, L. A., 2010. Environmental grants and regulations in strategic farm business decision-making: A case study of attitudinal behaviour in Scotland. *Land Use Policy*, 27, 415–23. https://doi.org/10.1016/j.landusepol.2009.06.003.

Thompson, P. B., 2010. *The Agrarian Vision: Sustainability and Environmental Ethics*. University Press of Kentucky, Lexington, KY.

Vukosavljević, S., 1983. *Istorija seljačkog društva III. Sociologija seljačkih radova*, in: Lukić, R. (Ed.), (Posebna izdanja, knjiga DXLVII, Odeljenje društvenih nauka, knjiga 89). Srpska akademija nauka i umetnosti, Belgrade.

Wallman, S., (Ed.) 1979. *The Social Anthropology of Work*. Academic Press, London.

12. Failure to Hive
A Co-narrated Story of a Failed Social Co-operative from the Hungarian Countryside

Éva Mihalovics and Zsüli Fehér[1]

Fig. 1. Sunset in Nagypatak. Photogaph by Lujza Nényei (2016).

My co-author Zsüli and I both believe that telling a story of and from the 'ground' is important. She sees this book chapter as a way to convey her ideas to new audiences, and to make them heard.[2] I see this story

1 Apart from ERSTE SEEDS and BADUR Foundation, Lujza Nényei, and me, all names of people and places are pseudonyms.

2 We plan to share this text with ERSTE SEEDS, Badur Foundation, and Zsüli's mentor as well.

 https://doi.org/10.11647/OBP.0244.12

as one that shows how much people's lives, and the decisions that they can make about their lives, are embedded in and defined by broader social and (onto)epistemic contexts (see Blaser, 2010; Koobak and Marling, 2014; Tlostanova, 2015, etc.). I approach this story not as a case study, but as a personal piece of situated and partial truth (Haraway 1988), translated[3] and filtered into the situated and partial knowledge production of academia on rural Hungary.

The structure of the chapter is rather unconventional as we present our narratives side-by-side: I give my interpretation of events and issues along with the story of rural development that Zsüli shares.

Zsüli does not speak English and does not "speak academia". I certainly don't speak the realities of Nagypatak as she does. At the end of the day, co-authoring this chapter means that I am the one with the opportunity and responsibility to translate someone's life into an academic text; to give an interpretation of her story. This inherent power imbalance comes with potential tensions and conflicts: co-authoring is not easy. To overcome at least some of these difficulties, Zsüli and I chose to write our chapter in a rather conversational style, bringing the academic and the everyday registers closer together. After several discussions in person, we moved online. Zsüli wrote the parts of the story that she found the most important, which I then translated and asked questions about. We have not agreed on everything, and we interpret certain events and issues differently. We have decided to leave these non-agreements, questions, hesitations, and even frustrations in the text, as if we were just talking. Talking about life.

To start our story, we need to locate Zsüli's tiny home village, Nagypatak, in contemporary Hungary. Nagypatak, with about 360 inhabitants, is situated by a river, under the Zemplén Mountains, in the north-eastern part of the country. This region is one of the twenty poorest in the EU (Eurostat, 2017), and the majority of people living in deep poverty there are Roma.

From 2010, Hungary has become an electoral autocracy (Ágh, 2015), or as Prime Minister Viktor Orbán referred to his own right-wing, populist regime in 2014, an "illiberal democracy" (for details see Bánkúti et al., 2012; Fekete, 2016; Bozóki and Hegedűs, 2018; and Chapter 1,

3 The decision to write the chapter in this form was inspired by decolonial and feminist authors such as Richa Nagar (2014) and Marisol de la Cadena (2015).

Kovács and Pataki). The realisation of this 'illiberal' democracy has been accompanied by the experience of near-constant attacks against the 'leftist liberal elite' by official and unofficially aligned government spokespeople and sources. This 'leftist' elite of Hungary has included the (internationally funded) NGO sector, academia and any independent media that has not been taken over by the Orbán regime's propaganda enterprises.

The government regime also states that its society is to be work-based and categorises its citizens as either deserving or undeserving (see Gans, 1993). Instead of granting broad welfare measures, it introduced a so-called public work scheme (*közmunka*). The slogan is "we provide work instead of social benefit" (*'segély helyett munkát adunk'*; see Csobai, 2020). This means that particularly in disadvantaged regions, often the only employer is the local municipality, which employs locals for shorter or longer periods[4] in *közmunka* (for more detail and critique of *közmunka* see Szőke, 2015). Nagypatak is one of the settlements where there are essentially no other options available for stable, wage-labour employment other than *közmunka*.

Entering the Story

I first met Zsüli when I was volunteering for an NGO mainly working with children and families in Nagypatak. During those months, I was trying to find my place and role in the NGO and ended up helping out at their social enterprise project, a guesthouse, which had opened only a year before. Zsüli's kids were involved in the children's programme of the organisation, and Zsüli, with other members of her family, was also helping around the guesthouse.

This NGO is the initiative (or mission) of a devoted social worker from Budapest; while looking for disadvantaged places that needed some form of help, he found the village and its people suitable for a long-term project. The NGO, and its programme, is only one of the four or five empowerment, development, and integration projects that I witnessed while working in the village as a volunteer, and later when

4 The contracts are typically for three months. Apart from the available budget, the extension many times depends on the benevolence of the local mayor or the social capital of the *közmunkás*.

I returned as a social researcher. The initiatives I know of are either NGO projects, just like the one I had been volunteering for, or run by big charity organisations, or by the municipality. Many times, I felt that this region and the tiny village of Nagypatak, despite being one of the twenty poorest of the EU, is 'overproject-ed' or 'over-helped'. The question emerges: what could justify running several similar programmes in a settlement with only 360 inhabitants? Working with the same children, the same families? Whose interests do these projects serve? And what happens to locals' initiatives?

For me, Zsüli's story is partially about how local ideas are shaped, transformed, and even hijacked or exploited by the structure of the organisations and the funding system aiming to achieve 'development'. Several authors have pointed out the problematic nature of the development paradigm,[5] and in the Hungarian context Imre Kovách (2013) has written about it at length. As he claims, projects and programmes targeting the countryside often serve the interests and provide income for the members of the 'project-class', who are mainly middle-class, educated, white-collar intellectuals. I do not question the benevolence and good intentions of experts and project-professionals working in these rural development programmes, but I agree with Kovách that, instead of solving problems, these intermediaries tend to reproduce inequalities, often in different forms.

About Beginnings

Zsüli: It was around October 2014 when my husband decided to run for mayor of our small village, Nagypatak. At home, we talked a lot and came up with various ideas about how we could help the villagers to maintain a better quality of life. We were thinking about starting a civil association, foundation or something like that. Then my husband met someone who suggested that we try to start a social co-operative. We looked it up, and after several discussions with friends in the village, we decided that we had found what we were looking for. This form of enterprise has a social impact, and we believed that to be absolutely

5 Among others see Asher and Wainwright, 2019; Eija, 2016; Ferguson, 1990; Howell and Pierce, 2001; Li, 2007; McEwan, 2018; Mosse, 2005, etc.

Fig. 2. Reed Stacks by the village. One of the local social enterprise ideas was to potentially use this abundant natural reed for weaving baskets or furniture. Photograph by Lujza Nényei (2016).

important. There were also other aspects which we didn't know about, to begin with, whose significance became evident later on.

But back in 2014 we were enthusiastic, and we felt that this was a good idea. If my husband won the election and became the mayor, a social co-operative could help to improve the lives of the locals. If he didn't win, we could still work on that improvement. It was already clear in those days that the government wanted to decrease the number of people employed in *közmunka* and that small settlements, villages and towns were going to lose governmental funding. Now, five years later, I can see that our line of thinking was right. After finding the right path there followed a period that I now call the 'times of daydreaming'. We searched for information and talked a lot, as well as holding a lot of meetings with the small group of locals who joined us.

Those days it was only us, and no-one from outside the village, no-one from outside of our world. There was me, my husband and other people from Nagypatak. The people joining us were Roma, living in poverty with their families, and both me and my husband thought that we should, and that we could, do something to give them an opportunity to get a decent salary, a stable existence.

I think we had a lot of really good ideas. For instance, since we have a beautiful but sometimes dangerous river in the village, we were thinking about offering help to the municipality with flood control and prevention works. Then we had this idea of helping with the communal waste management. Then since there are a lot of reeds outside the village—remember, there's a river there—we had this idea of manufacturing furniture from it, or at first just baskets. I admit that maybe we were naïve at some points, but we had lots of ideas and we felt that we had the energy.

The fact that we didn't have any money to invest didn't bother us at that point. Somehow, we had trust in the system—we thought that if we had a truly good idea, we could apply for funding—and then nothing could stop us. None of us had any entrepreneurial or leadership experience but we were sure that we would be able to do this. Then came the day of the mayoral election and my husband lost by very few votes. This was a bit depressing, and we had to adjust our plans since we were sure that the re-elected mayor of the village would not want to cooperate with us. Then only a few weeks later in November 2014, with the help of a very nice lawyer, we officially founded our social co-operative. We had eight members, and I was elected as the chair of the co-operative.

Later, in March 2015 we received wonderful news: we had won a grant from the regional Job Centre. We were very motivated and began the work immediately. The idea with which we won was to cultivate a big enough plot in the village to grow vegetables and herbs. My family has a very large plot, running down to the riverbank, so we thought we could begin by cultivating that piece of land and, if it worked, continue with the gardens and plots of the other members of the co-operative. I'm very passionate about gardening, about growing our own vegetables and everything else we can, and I'm especially passionate about herbs. To be honest, I don't particularly enjoy hoeing during awfully hot summer days... but I love to watch plants grow. And I love picking herbs in the wild. That year I was on childcare allowance with my smallest, and the rest of the group were either early school leavers or out of employment, even *közmunka* (public work scheme). If I remember well, only my husband was employed. The grant was exactly for people like my colleagues in the co-operative, people with only a primary school education, or not even that. In a way, I'm one among them, for I only completed the 8th grade.[6] But I'm very curious and I like learning. I feel that not getting a high-school diploma was a big mistake, and I regret it. To tell the truth, I still haven't given up on the idea of going back to school to get that diploma.

Éva: After a few meetings in person at Zsüli's home and talking through the story of the co-operative I moved back to my hometown. During those weeks we decided to continue working on the book chapter separately, each of us thinking through and writing our own parts and bits of the story. Then Zsüli wrote her parts and shared them with me in an online document. After that there came a lot of phone conversations, which mainly involved me asking Zsüli "when you write xyz, do you mean yzy?", "Is it OK if I translate your words like this?" and "It seems that we won't agree on this, but I want to add my interpretation as well. So here it goes....". When dealing with this exact part of our text, I had several questions and a few conundrums.

One was about the fact that the social co-operative's members wanted to earn money: they wanted some form of employment, and it seemed that the ideal working life imagined by them involved earning wage labour. I asked Zsüli if they had considered working in a '*kaláka*'[7] as an option for helping each other. This would be understood as a local version of a community or solidarity economy (cf. Mihály 2017, Miller 2009 or Gibson 2009, Gibson and Graham 2013), with members cultivating their plots, growing vegetables, fruits, and herbs together, for the benefit of everyone. And Zsüli said that no, they were interested in making money, in starting a 'real' economic enterprise and in providing education and (later) income for the members of the co-operative. This made me wonder why people living literally from the soil would prefer wage labour to liberating themselves from the dependency cycles of the 'real' economy?

There's an aspect to this conundrum that I want to discuss in detail. On the surface, this could be a story of neoliberal-capitalist

6 Final year of the primary school in Hungary. (Note by Éva.)

7 A '*kaláka*' was a popular and informal circle of mutual help during the years of state socialism. A typical example was building one's house as a group effort, everyone participating in the process. Then the next time a different member of the circle would receive help from the *kaláka*. This version of mutual help is similar to the Local Exchange Trading Systems (LETS) concept (see Mihály 2017 or Sík 1988).

entrepreneurial thinking, where land and labour are resources for doing business, for a profit. But from another perspective, I claim that at the centre of this story—as we tell it—we see the harmful effects not of capitalism on the rural imagination, but of class and ethnic relations, of social status, and of different concepts and understandings of work. In my interpretation, the development of project options for the social co-operative demonstrates how ethnicity (being Roma) is interpreted as a class issue (being poor) and as a social status issue (being under-educated, not knowing how to work or live 'properly').

I agree with Kovai (2016), who states that the assumed relationship of the Roma with the soil, with agriculture, is an accentuated site for reproducing hierarchical differentiation between Roma and 'white Hungarian' populations of rural areas. In Kovai's (2012, 2016, 2017) and Horváth's (2008, 2012) understanding, after the transition and its attendant economic and social changes, the previous spatial and thus social arrangements of the Hungarian countryside (which were not at all necessarily fair or pleasant) were disrupted. They discuss in detail how the ambition to eliminate segregated settlements, and to improve the housing conditions of Roma by moving them from the outskirts of villages to the centres, altered the perceived social and spatial structure of the villages. These authors connect this disruption of the differentiation between the ethnic minority of Roma and white Hungarians to the rise and strengthening of extreme right-wing, racist movements and groups.

After the transition in 1989, in Nagypatak there was also a phase of moving people from the '*cigánysor*'[8] to the centre of the village. Additionally, I have found that the 2008 economic crisis and regime transition in 2010 under Orbán, with its introduction of the workfare-based scheme, had a similar effect on Zsüli and her husband's thinking. Apart from finding solutions to poverty, the social co-operative provided a tool to maintain the social structures and thus their own higher social status in the village.

As Kovai writes, land cultivation plays a significant role in the maintenance of social status:

8 'Gypsy row' — Hungarian term for the streets on the settlements' outskirts where poor Roma people live in very bad conditions. There is usually no concrete paving the streets, and no public utilities such as sewage, water, or electricity in these areas.

> the elite of the village, those in positions of power, including the civil actors, social workers and those employed [Roma] in day-labour, all seem to share the opinion that local Roma do not know the means and methods of agricultural production. But this isn't only a deficit of professional knowledge, rather, a lack of work ethics, which, in their opinion is fundamental either for the self-sustainability of the village or the capability of performing well in 'market'-based employment. In this concept, the Roma citizens of the village are the ones that need 'enabling' and education. This means that Roma are put in the inferior, to-be-disciplined position of a child (2016, p. 140)

When I asked Zsüli what their aim was with this programme, she said that apart from providing a stable income for the families, they wanted the participants (who were Roma) to adjust to the needs and practices of the labour market, to teach them effectively how they could "survive" in it. I argue that Zsüli and her husband's concern with the young Roma men not being able to follow a 'normal' daily schedule, that is, get up in time, go to work, produce effectively, and finish jobs on time, means that the co-operative's educational agenda should be analysed taking into account the broader social, class- and ethnicity-based relations in the settlement. I emphasise this aspect partially because I find within this a contradiction. Drawing from my decade-long field experiences, I believe that Roma people do know how to work, how to perform tough, physical jobs, and they don't lack a routine—but their daily routine can be different, and not recognisable to 'white Hungarians'. In this instance, we're talking about Roma people, whose families have been involved in agricultural day-labour (working mainly with raspberries and apricot) for years.[9] Why would this not count as 'proper' job experience?

For development or empowerment projects targeting the countryside it is also crucial to have a nuanced, detailed, localised, and contextualised understanding of social arrangements and relations—and the changes imposed in these relations by the programmes themselves, so that they may be able to achieve their goals. Stepping into the process are white, middle-class, typically outsider actors, who usually undertake this intricate and complicated task with a lack of self-reflexivity regarding

9 Working on my Ph.D. project in the village, I learnt from the local Roma that many of them worked as seasonal day-labourers to support their families financially. This pattern fits into what several of the Hungarian literature described, see Hamar 2014, 2016 or Cseres-Gergely-Molnár 2014.

their own ethnic and class positionality and how these affect the body of the countryside. The ERSTE SEEDS and the Badur Foundation, explored below, are examples of the importance of such reflexivity.

Zsüli: The little group of six showed up every morning at eight o'clock, waiting for their list of tasks. They were motivated and in good spirits. After a few days, when we were cleaning the plot and preparing the soil for seeding, we got very bad news. It turned out that though the Job Centre was going to pay for the salary of the men for the first nine months, we had to pay the wages to start with, and it was only during the following month that they would reimburse our costs. The problem was that, as I said earlier, we didn't have any capital. Where would we get the wages from? And in this region, with all of these people being either unemployed or working in the *közmunka* programme, all of these people living in poverty... what were the Job Centres and the people leading the Job Centres thinking? How would we be able to pay wages—or, as a matter of fact, anything? Still, we were kind of lucky since we began working on the plot before signing the contract. Actually, I had to go to the regional centre, Miskolc, to talk to the regional Job Centre as the one in the small neighbouring town didn't give us the right information—and as there was no signed contract yet, we could just leave the programme. So, in the end we felt devastated, demotivated, and disappointed, but at least we didn't have to pay a fine. My husband and I decided to pay the workers from our own money—and we didn't have much.

When trying to find solutions, I got in contact with different NGOs that offered services for social enterprises (like NESsT), but I could not convince them to support us with 800-900 000 HUF, which would have made it possible to pay the workers at the end of the first month. Although they could not help us financially, they were interested in our project. I also made contact with the official governmental organisation for entrepreneurial initiatives (OFA). I found this type of networking very useful and was sure that with the help of these organisations, there was a future for the co-operative.

But in spite of those hopes, the whole experience was shocking and frustrating. When I had to tell the young men, our colleagues, that we could not pay them, could not honour their commitment, seeing them lose hope and trust... it was devastating. I find it incredibly sad that organisations like ours face so many difficulties and obstacles when they want to improve the lives of locals.

NESsT, OFA, ERSTE SEEDS, BADUR Foundation... And a Marriage Falling Apart

Zsüli: Two years passed, and we still could not start any profitable economic activity. In addition, in 2015 my marriage began to fall apart. It was a very difficult time in my life, and after one and a half years of separation, we divorced. I was left completely alone with three underage kids and no help (my eldest was already financially independent and living on her own in a nearby town. I do have relatives in Nagypatak, but after my husband left us, he did not pay for anything for a period, and our income significantly decreased). Those were tough months mentally and emotionally. I wasn't sure about my abilities and capabilities anymore and also felt left alone with the social co-operative. The impression was that my partners in the enterprise didn't want to put any time or energy into the project, that they only wanted money. Money, and promptly at that. I was frustrated and lonely in a supposedly profitable economic organisation that had only cost money so far. The other members didn't seem to care—they were like ghosts. They didn't show up at meetings; they didn't answer my calls. I was close to giving up the co-operative.

Éva: This is another sad part of Zsüli's story. Left alone with three children, and with no financial help to buy the children clothes or to pay for winter fuelwood. These months Zsüli was employed by the municipality as a *közmunkás*, but her income wasn't enough to heat all the rooms of the house, so she and the children moved into one room. She was as strong and persistent as ever, but the failed gardening project, and the disappearance of members, broke her spirit. When talking through this period, I asked her if she thought that her being a single woman played any role in what was happening in and around the social co-operative. She said yes, because she had to learn how to convince men to take her seriously, how to work with them. Particularly after her husband left, this became even more difficult, or impossible. I interpret this to mean that her social status, because of her gender and changed marital status, deteriorated, and this had consequences for the possible roles she was 'allowed' to take on in the village from this time onward.

Zsüli: The only person who tried to support me in finding new ways forward for the co-operative was my mentor from OFA, who said that there were upcoming grant applications that he thought we had a good chance of winning. He said that we could try and apply, and if we didn't succeed, I could still liquidate the co-operative. Then, in February 2017, this friend and mentor of mine invited me to participate at an informational event held by ERSTE SEEDS.

Up until that point I had no idea who they were or what they were doing. But the co-operative had so many ideas, some of them must have been good ones... so, after the event the two of us decided that we should pick one and write a proper business plan for it and apply to the ERSTE SEEDS social enterprise incubation programme. At some point my mentor asked me how I felt about bees. He knew a very successful beekeeper in a neighbouring village, a family enterprise. They were producing high quality honey, mainly for export. They even delivered to Japan. If I remember well, my mentor had helped them with the first steps of building their enterprise, and they had been in touch after as well. I didn't know anything about the 'little buzzies' but found the idea exciting. I knew that bees were important for us and the world to survive, and I was keen to participate in a project with them. We talked to several beekeepers in the area and it seemed that producing equipment and tools for them could be a winning idea. The two of us worked a lot on the application that we submitted. Then, I was kind of shocked in a positive way when we learnt that from 202 applications, ours was amongst the sixty-eight selected.

Éva: ERSTE SEEDS is a social enterprise incubation programme of Erste Bank, which has been running for several years in Hungary. The chosen initiatives and organisations participate in an eighteen-month process to learn how to build up an enterprise, interrogate what 'social' could mean in a social enterprise, learn about marketing and risk management, and gain the opportunity to build up a strong business plan. At the end of the process, the projects are evaluated, and the best ones receive money, and an opportunity to pitch their projects to real investors.

As I participated in the programme, I have first-hand experience of how ERSTE SEEDS handled questions (or rather problems) of

Fig. 3. A few of George's hives, the successful apiary enterprise in a neighbouring village. Photograph by Zsüli Fehér (2018).

distance and income during the training period. I found that despite their benevolent intentions, the inherent, unquestioned neoliberal-capitalist agenda of the programme led to a certain blindness and lack of knowledge of the realities of poverty in rural Hungary. I was present as a volunteer of the NGO running a social enterprise guesthouse in Nagypatak. The training took place in the capital, Budapest, once every month. As a white, middle-class 'visiting volunteer' of the NGO, living in the suburbs of the capital, I didn't face any problems attending the course. But I came to learn that both my local colleague from the guesthouse, who was a young mother, and Zsüli had troubles paying for the train ticket (a full price return ticket costs about 8000HUF, and Zsüli's monthly income was 131 000HUF), and since this was a whole day-long trip and programme, they also faced childcare difficulties. I asked the organisers whether they were planning to pay for the tickets, and they seemed a bit surprised. They told me that they had not anticipated that many participants from such far-away, rural areas. I asked if it was possible to move at least some parts of the training closer to those faraway places that they had, after all, accepted applications from, and they said that they didn't think so. But, as a solution to the conundrum, they thought that they could reimburse a certain percentage of the travel costs for each participant. And this did happen. But reimbursement presumes that people had money to spend on the tickets to begin with, and that they would not miss this from their survival-oriented daily budget.

I was shocked to see how evidently the presence and ideas of higher-status and higher-income people already placed them at an advantage before the training had even begun. For me this suggests, apart from appropriating and reshaping local ideas and initiatives, that the system works in a spatially exclusionary way, narrowing down the chances of individuals from those faraway areas to participate, let alone 'catch up'.

I argue that this spatial determination and/or blindness of the bank's social responsibility programme should be analysed in connection with the strange structure of the Hungarian civil sector, which was captured by the intellectual elite of the capital city after the transition period (see Lomax, 1997; Hann, 1995). This means that there are many outreach programmes, initiatives planned and handled in the capital—and performed on the body of the countryside. Building on my field and volunteer experiences in the sector, I claim that scholarly knowledge production should investigate in detail the issues and assumptions around local-ness and grassroot-ness in the Hungarian context. It is important to find out who and how defines who and what counts as local and as grassroots in the Hungarian context? And with what agendas and interests in mind? For instance, from the four development projects that I am aware of which target the tiny village of Nagypatak, three were initiatives from the 'outside'. One of them was the mission of the social worker I was a volunteer for, and two others were run by an NGO and a big charity organisation where the programme leaders did not live in the village, nor even in the region, and only visited their 'worksites'. What I found perplexing was that these projects tended to see themselves as 'local', and one of them even claimed to be 'grassroots'. This latter project justified its self-definition with the fact that the organisation was one of the 'independent', 'progressive' NGOs that found itself under attack from the government.

Thus, in their understanding, being outside of the financial and institutional regulatory and support system of the extreme right-wing government guaranteed a positive label or signified that they were of the 'local' and belonged to wider society. I strongly argue that from the side of NGOs this 'myth of local-ness' (my term), and the need to be defined and credited as 'local', should be analysed and contextualised through nuanced ethnographies, paying attention to the spatial aspects of the issue.[10] I also argue that without an analysis of this kind, we won't get a better understanding of why and how the countryside remained silent and indifferent when the government attacked the NGO sector.

When projects and programmes *are* initiated by local intellectual elites (as in Kovai's example, by the mayor or, in other cases, by social workers and teachers), they do not pay attention to their own class, ethnicity, or gender, and the risks and consequences of these positionalities. How can initiatives of local elites, making decisions about projects, applying for funding then distributing the funding, claim that they build their enterprises on the basis of democratic decision-making and that they approach their beneficiaries as equal partners (see Mihály, 2019)? Pre-existing individual and broader relations, such as ethnic, class, gender, and other determinants of social status are involved in the complicated social context they aim to tackle. To be able to detect the "manipulation of the local elites" (Mosse, 2005: 5) and address the impact of such manipulation are key tasks for rural development projects.

What part do international donor organisations play in this issue? These donor organisations typically claim to assist the 'thriving' of local initiatives while being ignorant of or insensitive to local contexts such as ethnicity, class, etc., or the accessibility of their programmes, and tend to not recognise (or even to ignore) their own role in the constant reproduction of local hierarchies and inequalities.

I want to offer a way to approach this 'spatial blindspot'. Decolonial scholars working on the realities of eastern European (semi)peripheries claim that this region serves as Europe's inner 'Other' (or deploying Boatcă's (2007) expression: Europe's "pathological region"), and is

10 In the vein of the works of Forbes 1999, Howell and Pierce 2001, Butcher 2007, Leve 2207, Karim 2008, etc.

in constant need of development in order to catch up with the West. But these authors also claim that the region's societies' intellectual elite follow, by default, a Western neoliberal-development agenda which means that they effectively commit epistemic self-colonisation (cf. Tlostanova, 2015; Koobak and Marling, 2014; Kiossev, 1999; Melegh, 2006, etc.). To understand this tendency of self-colonisation, we must take into consideration that after the transition many of our intellectual elite were (and are) trained and educated (either formally, obtaining university degrees at Western universities, or informally, through training) and funded (through the NGO projects they work in) by the neoliberal West. Thus, I claim that our rural (semi)peripheries can be understood as targeted 'Others' of and for intellectual elites from the centre, who operationalise the 'myth of the local'—while remaining 'Others' to western Europeans, the EU, and international donor organisations.

Zsüli: At this point the social co-operative was mainly me and a young man from Nagypatak who participated in the training once or twice. But it was mainly me. During these months I learnt a lot. I learnt how difficult and complicated it is to run even a very simple and small business. It requires a lot of preparation and research. We had to think over a lot of things to find out if an idea made sense and was doable or not. I'd say that during these months I realised how naïve we were at the beginning. We didn't get any money in the end, and didn't find an investor, but I found it a great opportunity for networking—and built some true friendships as well. Now I feel that this was the best time of my social co-operative period.

In the end our co-operative and our idea of producing beekeeping equipment didn't make it to the final round, and nor did we get any money or meet an investor. So, my life went back to what I call 'normal'. Then, in spring 2018, it seemed that there was a chance of a breakthrough. My mentor did not forget the social co-operative, and he tried to find a way to help us. Or, at that point, to help me, as there were no other active members working or at least thinking about the future of the organisation. My mentor called with good news: he had shared the story of the co-operative with the Badur Foundation,[11] and they had decided to support a pilot project of ours.

Fig. 4. Zsüli in gear working with the hives at George's apiary. Photograph by Zsüli Fehér (2018).

We talked through our options, and we thought that based on the lesson we had learnt from the ERSTE training, we would try something smaller, something easier. I was still in love with the little buzzies so a small

11 From amongst the organisations that play a significant role in Zsüli's story, OFA is a governmental body, ERSTE SEEDS is a social responsibility initiative of Erste Bank, and NESsT and Badur are international donor agencies, working in several countries, that finance 'local' development projects.

apiary enterprise seemed to be a nice and logical decision. We got money from the Badur Foundation to learn from the best apiary in the area, to buy a starter kit (hives and protective gear), and a few bee families. The final goal was still to produce apiary equipment and tools, because we were sure that it would be a good and profitable idea. I had to find new people to join the co-operative and participate in the project. Two people agreed to join me to learn about keeping the little buzzies. Once again, I felt lucky and motivated. George, the successful beekeeper who taught us, was a very good teacher. He explained and showed us everything in an easy-to-follow way. I learnt a lot really quickly. Beekeeping seemed to me a wonderful profession. Then early in the summer one of the two participants who joined said that she needed to quit because of mental problems. Then only a few weeks later the other new member left as well. I was disappointed and frustrated. I could not understand why these people didn't see beekeeping as a chance, why they didn't think about it and the co-operative the way I did. Once again, I was left alone, and I could not find anyone in the village who was interested either in this project or in the social co-operative. By October it was clear that this, again, was going to fail.

The End...

We sat down with the Badur Foundation to talk through the possibilities. I told them that I saw no chance of continuing. They were very nice and asked me to wait till February and to try and find new members for the co-operative. But by that time, I was through... I had talked to several people, tried to convince them, but I found no one who wanted to join the co-operative, to put in the effort, time, and yes, sometimes money, to build up a social enterprise. There was no energy or ambition left, I was tired physically, emotionally, and mentally as well. I was frustrated, disappointed, and to be honest, I felt betrayed by the people whom I had wanted to help gain a chance to live a better and more stable life, but who didn't show any interest. I got a letter from the Court of Company Registration, claiming that I had to modify the charter of the organisation, and that there should be new members joining the co-operative, either a municipality or a charity organisation. If we failed to follow these recommendations, the co-operative would be legally terminated. I thought that this was the easiest way to get rid of the co-operative: by doing nothing. The Badur Foundation accepted my decision about giving up on the apiary project, and I transferred their money back.

Éva: I remember that I was already in Durham and checking on Zsüli by phone when she told me that she was about to close the apiary project, and shut down the whole co-operative. I understood her reasons, and in particular the fact that she didn't have more money, time, or energy to spend on the project. I also knew that winter was approaching again and that she had to prioritise the needs of her family.

On another level I felt frustrated, and even angry. Here was this persistent, hard-working woman, full of energy and ideas and despite all this, her life was not getting any better, and social enterprises and development projects could not help or solve her problems. I also felt that I, as a volunteer of that NGO, the NGO itself, the numerous projects in the village, the mentor and his approach, the funding and supporting system available for rural initiatives, government grants like the one provided by the Job Centre or the ERSTE incubation programme and the Badur Foundation's pilot-project—none of us necessarily 'helped' people living in the Hungarian countryside. Instead, though not intentionally, as part of the broader social and global structural violence (see Farmer 1996), we reproduced and maintained inequalities.

In my understanding the NGOs under attack from the right-wing populist Hungarian government and the international donors allocating their funding are all part of the same development arena. This arena is an imagined but at the same time real site of encounters for different parties with different interests, agendas, and realities. These differences are not necessarily the products of different cultures, practices or customs, but as Blaser (2009, 2010, 2014) points out, often they are the materialisations of different worlds around us—different ontologies and thus epistemologies.

For me, as a social science researcher, Zsüli and the co-operative's story is about different aspects of local-ness. First, I argue that development projects targeting the post-state-socialist countryside need to be aware of local ethnic, class, gender, and other potentially important relations and contexts, and, moreover, that all actors in the development arena need to be reflexive and accountable for their own roles in maintaining or changing the status quo. Further, we need nuanced and detailed, long-term ethnographic accounts, localised-contextualised understandings of development projects targeting the Hungarian countryside, in the vein of the works of Tania Li (2007) or David Mosse (2005). Post-state-socialist rurality is a suitable site for deploying a decolonial lens, as such an approach allows the different ontologies at play between the NGO sector, donor agencies, and academic knowledge production circuits within these projects to become visible.

Part of my present task is to gain a deeper understanding of the 'myth of local-ness', how and why this label is deemed important and is enacted by different actors participating in development projects that target the Hungarian countryside. The leadership of the civil sector by the intellectual elite, and my field experiences and research, suggest that this leadership form has long-term consequences. It shapes, defines, and channels what counts as a promising 'local' idea (or ignores and denies initiatives that do not fit to its scheme). I claim that a critical and self-reflexive analysis of this agenda needs to be undertaken based on the experiences recounted above, exploring our potential epistemic self-violence within this system.

These are hard times for the leftist, progressive intellectual elite, its academia, and its NGOs. However, the volunteers of NGOs, and often we, those who produce knowledge on the Hungarian countryside, are as much a part of the problem as the solution.

References

Ágh, A., 2015. De-Europeanization and De-Democratization Trends in ECE: From The Potemkin Democracy to the Elected Autocracy In Hungary. *Journal of Comparative Politics*, 8(2), 4–26.

Asher, K. and Wainwright, J., 2019. After Post-Development: On Capitalism, Difference, and Representation. *Antipode*, 51, 25–44. https://doi.org/10.1111/anti.12430.

Bánkuti, M., Halmai, G. and Scheppele, K. L., 2012. Hungary's Illiberal Turn: Disabling the Constitution. *Journal of Democracy*, 23(3), 138–46. https://doi.org/10.1353/jod.2012.0054.

Blaser, M., 2014. Ontology and indigeneity: on the political ontology of heterogenous assemblages. *cultural geographies*, 21(1), 49–58. https://doi.org/10.1177/1474474012462534.

——, 2010. *Storytelling Globalization from the Chaco and Beyond.* Duke University Press, Durham, NC and London. https://doi.org/10.1515/9780822391180.

——, 2009. The Threat of the Yrmo: The Political Ontology of a Sustainable Hunting Program. *American Anthropologist,* 111, 10–20. https://doi.org/10.1111/j.1548-1433.2009.01073.x.

Boatcă, Manuela, 2007. The Eastern Margins Of Empire. *Cultural Studies,* 21(2–3), 368–84. https://doi.org/10.1080/09502380601162571.

Bozóki, A. and Hegedűs, D., 2018. An externally constrained hybrid regime: Hungary in the European Union. *Democratization,* 25(7), 1173–89. https://doi.org/10.1080/13510347.2018.1455664.

Butcher, J., 2007. *Ecotourism, NGOs and Development: A Critical Analysis.* Routledge, London and New York. https://doi.org/10.4324/9780203962077.

Cadena, M. de la, 2015. *Earth Beings: Ecologies of Practice across Andean Worlds.* Duke University Press, Durham, NC. https://doi.org/10.1515/9780822375265.

Cseres-Gergely Zsombor — Molnár György, 2014. Közmunka, segélyezés, elsődleges és másodlagos munkaerőpiac, in: Kolosi, Tamás, Tóth, István György, (szerk.). *Társadalmi Riport 2014.* TÁRKI, Budapest, pp. 204–25.

Csoba, J., 2020. *Revitalisation of the Household Economy: Social Integration Strategies in Disadvantaged Rural Areas of Hungary.* Series: Prekarisierung und soziale Entkopplung — transdisziplinäre Studien. Springer VS, Wiesbaden. https://doi.org/10.1007/978-3-658-29350-5.

Eija, M. R., 2016. Toward a Decolonial Alternative to Development? The Emergence and Shortcomings of *Vivir Bien* as State Policy in Bolivia in the Era of Globalization. *Globalizations,* 13(4), 425–39. https://doi.org/10.1080/14747731.2016.1141596.

European Commission, 2018. Civil society monitoring report on implementation of the national Roma integration strategy in Hungary. Focusing on structural and horizontal preconditions for successful implementation of the strategy. Brussels. https://doi.org/10.2838/034974.

Farmer, P., 1996. On Suffering and Structural Violence: A View from Below. *Daedalus,* 125(1), 261–83. https://www.jstor.org/stable/20027362.

Fekete, L., 2016. Hungary: Power, punishment and the 'Christian-national idea.' *Race & Class,* 57(4), 39–53. https://doi.org/10.1177/0306396815624607.

Ferguson, J., 1990. *The Anti-Politics Machine: "Development," Depoliticization, and Bureaucratic Power in Lesotho.* Cambridge University Press, Cambridge.

Forbes, A., 1999. The importance of being local: Villagers, NGOs, and the world bank in the Arun valley, Nepal. *Identities Global Studies in Culture and Power,* 6(2–3), 319–44. https://doi.org/10.1080/1070289X.1999.9962647.

Gans, H., 1993. Mire szolgálnak az érdemtelen szegények? *Szociológiai Szemle,* 3–4.

Ganti, T., 2014. Neoliberalism. *Annual Review of Anthropology*. 43, 89–104. https://doi.org/10.1146/annurev-anthro-092412-155528.

Gibson, K., 2009. The Community Economies Collective. Building community-based social enterprises in the Philippines: diverse development pathways, in: Ash A. (ed.): *The Social Economy: International Perspectives on Economic Solidarity*. Zed Books, London, New York, pp. 116–38.

Gibson-Graham, J. K., Cameron, J., Healy, S., 2013. *Take Back the Economy. An Ethical Guide for Transforming our Communities*. University of Minnesota Press: Minneapolis, MN. http://dx.doi.org/10.5749/minnesota/9780816676064.001.0001.

Hamar, A., 2016. Idénymunka a zöldség-gyümölcs ágazatban, in: Kovács, K (ed.). *Földből élők: Polarizáció a magyar vidéken*. Argumentum Kiadó, Budapest, Magyarország, pp. 97–115.

——, 2014. „Mi lenne nélkülük...?!" Külföldi és haza napszámosok egy Dél-Pest megyei kistérségben, in: Nagy, Erika, Nagy, Gábor (szerk.). *Polarizáció–Függőség–Krízis: eltérő térbeli válaszok*. MTA KRTK: Békéscsaba, pp. 18–27.

Hann, C., 1995. Subverting Strong States: The Dialectics of Social Engineering in Hungary and Turkey. *Daedalus*, 124(2), 133–53. https://www.jstor.org/stable/20027300.

Haraway, D., 1988. Situated Knowledges: The Science Question in Feminism and the Privilege of Partial Perspective. *Feminist Studies*, 14, 575–99. https://doi.org/10.4324/9781003001201-36.

Horváth, K., 2012. Silencing and Naming the Difference, in: Stewart, M. (ed). *The Gypsy 'Menace'. Populism and the New Anti-Gypsy Politics*. C. Hurst and Co., London, pp. 117–37.

——, 2008. „A cigány különbségtétel." http://cieh-programmes.univ-paris3.fr/docs/Minorite/1hadastol.pdf.

Howell, J., and Pearce, J., 2001. *Civil Society & Development: A Critical Exploration*. L. Rienner Publishers, Boulder, CO.

Karim L., 2008. Demystifying Micro-Credit: The Grameen Bank, NGOs, and Neoliberalism in Bangladesh. *Cultural Dynamics*, 20(1), 5–29. https://doi.org/10.1177/0921374007088053.

Kiossev, A., 1999. Notes on Self-Colonising Cultures, in: *After the Wall: Art and Culture in Post-Communist Europe*. Modern Museum, Stockholm. http://monumenttotransformation.org/atlas-of-transformation/html/s/self-colonization/the-self-colonizing-metaphor-alexander-kiossev.html.

Koobak, R. and Raili, M., 2014. The decolonial challenge: Framing post-socialist Central and Eastern Europe within transnational feminist studies. *European Journal of Women's Studies*, 21(4), 330–43. https://doi.org/10.1177/1350506814542882.

Kovách, I., 2013. A felemelkedő projektosztály, in: Czibere, I., Kovách, I. (Eds.), *Fejlesztéspolitika — vidékfejlesztés*. Debreceni Egyetemi Kiadó, Debrecen, pp. 157–69.

Kovai, C., 2017. *A cigány-magyar különbségtétel és a rokonság*. L'Harmattan Kiadó, Budapest.

——, 2016. Önellátó függőség: A közfoglalkoztatás társadalmi beágyazottsága egy Tolna megyei faluban, in: Kovács K (ed.). *Földből élők: Polarizáció a magyar vidéken*. Argumentum Kiadó, Budapest, pp. 180–96.

——, 2012. Hidden Potentials in 'Naming the Gypsy': The Transformation of the Gypsy-Hungarian Distinction, in: Stewart, M. (ed). *The Gypsy 'Menace'. Populism and the New Anti-Gypsy Politics*. C. Hurst and Co., London, pp. 281–95.

Leve, L., 2007. 'Failed Development' and Rural Revolution in Nepal: Rethinking Subaltern Consciousness and Women's Empowerment. *Anthropological Quarterly*, 80(1), 127–72. https://www.jstor.org/stable/4150946.

Li, T., 2007. *The Will to Improve: Governmentality, Development, and the Practice of Politics*. Duke University Press, Durham, NC: Duke University Press.

Lomax, B., 1997. The Strange Death of 'Civil Society' in Post-Communist Hungary. *Journal of Communist Studies and Transition Politics*, 13(1), 41–63. https://doi.org/10.1080/13523279708415331.

Lubarda, B., 2019. Far-right agricultural alternatives to right-wing populism in Hungary: the "real" Caretakers of the Blood and Soil. *Culture della Sostenibilità*, 24, 319–35. https://doi.org/10.7402/CdS.24.002.

Masterson, T., Kawano, E., and Teller-Elsberg, J., 2010. Solidarity Economy I: Building Alternatives for People and Planet: Papers and Reports from the U.S. Forum on the Solidarity Economy 2009. Center for Popular Economics, Amherst, MA.

McEwan, C., 2018. *Postcolonialism, Decoloniality and Development* (2nd ed.). Routledge, London and New York. https://doi.org/10.4324/9781315178363.

Melegh, A., 2006. *On the East–West Slope: Globalization, Nationalism, Racism and Discourses on Central and Eastern Europe*. CEU Press, Budapest and New York.

Mihály, M., 2019. Újratermelődő "gettók"? — A helyi fejlesztés lehetőségei és korlátai egy szélsőségesen marginalizált kistelepülésen Reproducing "ghettos"? — Development in a village undergoing advanced peripheralization. *Tér és Társadalom*, 33/4, 132–56.

——, 2017. Mit értünk társadalmi vállalkozás alatt és miért kutatjuk? — Narratívák a nemzetközi szakirodalombó. *Eszak-magyarorszagi Strategiai Fuzetek, Faculty of Economics, University of Miskolc*, 14(1), 101–15.

Miller, E., 2009. Solidarity Economy: Key Concepts and Issues, in: Kawano, E., Masterson, T. N., Teller-Elsberg, J. (Eds.), *Solidarity Economy I. Building Alternatives for People and Planet; Papers and Reports from the 2009 U.S. Forum on the Solidarity Economy*. Center for Popular Economics, Amherst, MA, pp. 1–12

Mosse, D., 2005. *Cultivating Development: An Ethnography of Aid Policy and Practice*. Pluto Press, London and Ann Arbor, MI. https://doi.org/10.2307/j.ctt18fs4st.

Nagar, R., 2014. *Muddying the Waters. Coauthoring Feminisms across Scholarship and Activism*. University of Illinois Press, Urbana, Chicago, and Springfield. http://dx.doi.org/10.5406/illinois/9780252038792.001.0001.

Ong, A., 2006. *Neoliberalism as Exception: Mutations in Citizenship and Sovereignty*. Duke University Press, Durham, N.C. http://dx.doi.org/10.1215/9780822387879.

Sík, E., 1988. *Az "örök" kaláka — A társadalmi munka szociológiája*. Gondolat Kiadó, Budapest.

Szőke, A., 2015. A 'Road to Work'? The reworking of deservedness, social citizenship and public work programmes in rural Hungary. *Citizenship Studies*, 19(6–7), 734–50. https://doi.org/10.1080/13621025.2015.1053790.

Tlostanova, Madina, 2015. Can the post-Soviet think? On coloniality of knowledge, external imperial and double colonial difference. *Intersections*, 1(2), 38–58. https://doi.org/10.17356/ieejsp.v1i2.38.

Concluding Thoughts

The Contributors

As a collective, we want to end this volume with a shared reflection on two key points: the place of 'critical' social science scholarship in the eastern European region, and an outline of some of the challenges it faces.

There is growing attention to recent developments in eastern Europe in terms of the region's governance and government continuities with socialism, most notably some countries' (re-)turn to authoritarianism, or similar hybrid models (Bugarič, 2019; Krastev, 2018). The notion that the region was ever transitioning to be like the West from the depths of state-socialist repressions and inefficiencies has been thoroughly problematised for its assumptions around the inevitability and path-dependent linearity of such a development trajectory, with the West setting the terms for what counts as 'progression'. At the same time, any so-called transition to capitalism may be said to be well and truly over, as EE states have entirely neoliberalised their economies and public sectors, with political leaders keeping the spectre of and nostalgia for the socialist past alive as a "boogeyman"—as a threat that the socialist past could return—and thus using evolving interpretations of history as "disciplinary devices" (Chelcea and Druţă, 2016; Nadkarni, 2020).

The current populist, authoritarian turn in the region threatens democratic practices and foundations, including the terms on which any civil society may function. The forms and activities of the civil society sector underpin several chapters in this volume. Heightened antagonism from governments and state agencies towards civil groups may be part of a blanket policy of repression (as in Hungary, see Kovács and Pataki, Chapter 1; or for Poland see Szulecka and Szulecki, 2019), or specific to the environment sector, particular to identifiable protests or causes,

 https://doi.org/10.11647/OBP.0244.13

such as when environmental groups agitate to threaten or undermine a country's energy policy (for some consideration of these relations in Czechia, see Novák, Chapter 2; and Černik, Chapter 3). *Where* ideas around environmentalism or conservation come from matters, as do the methods by which they are realised, because they influence local buy-in and perceived legitimacy (see Černik, Chapter 3; Hrckova, Chapter 4; Lubarda, Chapter 5; Püsök, Chapter 7; Iordăchescu, Chapter 8; Blumberg, Chapter 10; for (in)ability to 'participate' see Mihalovics, Chapter 12), and thus have significant consequences both for environmentalism and the vibrance (of participation or donor support) in any emerging 'third' sector.

As researchers, it can be challenging to write about eastern Europe and the political trajectory of some of its states without getting stuck on questions concerning ideology, and where researchers lie in relation to state-promulgated values and programmes. As reflective and engaged researchers, we need to consider the accusations of subjective bias that have come to plague the social sciences, which suggest that researchers' political bias and personal circumstances are thought to influence analyses and research results.

***

Many of the contributors of this edited book self-identify as local or 'native' scholars. We live amongst those we write about, where our research 'field' is also the place we call 'home', and the context where we take up active political roles to better understand and to elevate and represent particular causes and voices. The earlier chapters of this volume are the result of years of activism on behalf of their authors—Mikulás Černik in Chapter 3 terms his research process for this engagement 'militant ethnography'—with most other chapters also engaging in and using the tools of participant observation. This approach requires long-term embeddedness and an emphasis on active participation. Together, we posit that it is important to wear these activities and commitments 'on our sleeves', and engaged ethnography 'about home' makes our research rigorous as well as epistemologically unique.

We are all in the challenging position of conducting research that at times can be read as critical of mainstream structures of power. Yet, we do not conflate disagreement with the values of governments or their policies (which may motivate research) with the research results, which may well be critical of those values. Critical research often investigates the provenance and promulgation of societal narratives, considering how they are mobilised as potential social engineering tools. Beyond the underpinnings and operations of state agendas, a strong regional research focus lies in the question of how state institutions fail to live up to (or institutionalise) the rule of law. Across eastern European contexts, state function decline is represented through the hollowing out of due process, legal regularity and predictability, including judicial independence and governmental transparency. Tied into these processes are myriad social consequences, not least the concretisation and (re-)normalisation of nepotistic pathways—who you know and how much money you have—to "live and get by" (Saitta et al., 2013). These processes can be viewed through the evolving prisms of the state-citizen relationship, and the individual's place within society.

These negotiations are not in any way confined to eastern Europe. Neoliberal models of agency make the individual responsible for everything (cf. Watts, 1994)—for themselves through their successes, as well as their downfalls. The individualisation of responsibility includes action around climate change and biodiversity loss, where these processes are not recognised as the outcomes of systems failure, as market forces are disembedded from social institutions (the Polányian critique of market society). With the collapse of socialism and the erosion of Keynesian welfare states, we would argue that there is a risk that we are all potentially headed in an increasingly authoritarian-neoliberal-oligarchic direction (Bohle and Greskovits, 2019).

Critical research that is attentive to the machinations of these 'macro' trends around state-prescribed individual responsibilities and rights, and for how these intersect with local realities and needs, is vital in this space and in this moment. Some of the many environmental consequences of this neoliberal-authoritarian turn have been canvassed and interrogated by chapter contributions to this book. Resource governance forms today share many parallels with the access politics of socialist times, with the elite and well-placed able to capitalise on their social networks to gain ownership and control. The region is replete with examples of local resources being controlled through mafia-like fiefdoms that reassert local hierarchies and contested roles and forms of the state (Thelen et al., 2011): from fisheries, forests, and land, to agricultural subsidies (Dorondel, 2016; Gonda, 2019; Kovács, 2015; Schmidt and Theesfeld, 2012; Stahl, 2010; Vasile, 2019). Despite the entry of green discourses and pressures for better environmental management, environmental conditions have deteriorated throughout the region over the past three decades. The proliferation of private actors interested in extraction and profits, the increased mechanisation of all resource sectors, a panoply of enormous infrastructure and development projects majority-financed by the EU, and so on, have taken their toll. In addition, accession to the EU has meant the wholesale adoption across the continent of the most harmful environmental policy, the CAP, which has contributed to huge biodiversity losses and degradation in farmland quality, as the policy incentivises the intensification and mechanisation of agricultural activities (Mihók et al., 2017).

Various contributions to this volume make strong cases for how sustainable socio-environmental futures are threatened by local interpretations or realities of displacement and marginalisation (see Coțofană, Chapter 6; Püsök, Chapter 7; Iordăchescu, Chapter 8; for consideration of working landscapes with little space for local capital or practices see Brawner, Chapter 9; Dikovic, Chapter 11 and Mihalovics, Chapter 12 focus on differential local meanings and identities as they intersect with rural development). Yet, unsurprisingly, such research is amongst the first to be diminished. The relative lack of accessible, evidence-based internal societal critique in eastern Europe is a reflection, we argue, of the emerging public milieu, where even when outright (oppositional) critique may be found, there is no corresponding public movement to take it forward.

From the eastern European region, the most well-known attacks on independent thought and research 'from within' come from Hungary, as best illustrated through the fate of the Central European University in Budapest. In addition to this, the Hungarian government has mandated the removal of gender studies as a recognised degree, and by taking education and research institutions across the country (including the Hungarian Academy of Sciences) under government control, has redirected research fund bids. These recent developments give rise to serious concerns around the independence of the academic sector, as the relationship and service of research seems re-targeted to aid private actors. These changes also limit the scope and possibility of critical socio-political topics that might receive funding and inquiry.

A further reason for diminishing (or stagnant) quantity of critical social research are the realities of academic (and post-graduate) life within universities and research departments. Universities in EE, following the example of universities elsewhere, have been gradually neoliberalised over the past two decades. Universities today function as increasingly privatised, competitive business enterprises, as they are recast and expected to work in the state's interests and contribute to its economic productivity—and cater to students who are classified as 'paying customers'. The pressure to churn out degrees and focus on individual career achievements has come at the expense of rigorous, 'slow' scholarship and scholarship practiced as a community of participants. This is because resources (time and finances) for long-term, engaged action research are all the more difficult when required to fit alongside precarious teaching contracts, 300 students, and a lack of research funds. From several of our own experiences, many universities in the EE region fail to be meritocratic, as they continue to operate and reward scholarships and positions via nepotistic routes that speak to the ingrained hierarchies of the sector. There are also serious questions to pose and answer around the academic sector's role in the emergence and consolidation of contemporary power hierarchies in the region. Western academic models also need to be questioned and not necessarily emulated, as these education systems promote a self-centred, highly individualistic career trajectory that is non-cooperative and competitive, wherein hierarchy and attendant servility also pervade career pathways. The current academic status quo diminishes the actions and outcomes of research as well as early career researchers, as those with the 'right' connections and personalities to compete are again and again privileged and promoted, with many doctoral and graduate students falling by the wayside.

As scholars, it is furthermore important to continuously interrogate how we construct our critique. Scholarship and conversations around 'decolonising knowledge' recognise the eastern European region's imperial/colonial past (Tlostanova, 2019), and highlight the difficulty of providing critique through language and through the views of those about whom we write, without recourse to the same canonised

Western orthodox theoretical frameworks and dogma. We find affinities with decolonial movements advocating to break with "familiar citational infrastructure" (Nagar, 2019: 5) and to submerge ourselves into the worlds of those about whom we write, and the values and things that are worth fighting for and preserving. In many ways, such an ethics and research mode presupposes "a shared hunger for an intense transformative engagement with social worlds that can inspire intellectual and political agitation by remaking how we locate ourselves in relation to the bodies, battles, wisdoms, and worlds among which we move, and that we represent and reimagine," (ibid.: 22). To realise such novel positionalities and epistemological insights requires changes not only to our academic practice 'in the workplace', but also to our everyday lives, which must also embody solidarity. In this volume, such solidarity is solidified through the commitment to gather together a group of scholars *from* the region, to write *about* and *for* this region first and foremost, but also with the intention to reach colleagues and friends *beyond* it, in the hopes that the issues presented here resonate with other places. This commitment has thoroughly motivated our decision to publish with Open Book Publishers, particularly in light of widespread difficulties around research access (arising from institutional and financial factors) experienced by local scholars.

Perhaps due to our lives 'on the ground', from within the EE region, the prevailing winds are that of pessimism about the immediate future and about the possibilities for imagining, let alone realising, alternative and more hopeful futures. There has been a recent turn within geographical research to emphasise hope—or "emancipation" (Scoones et al., 2018)—as lights at the end of (or alongside) tunnels of authoritarian practices. These reinscribe somewhat linear expectations and ideas about progressivism, using a language and framework not grounded in most of our interlocuters' prisms and worldviews, let alone our own personal experiences. However, the only way to understand the forms and possibilities of hope being realised is to engage and build wide-ranging communities and communities of practice, where researchers must act as cultural translators in the service of an environmental cause that is borderless and shared. This is where we find hope and situate our activist scholarship.

References

Bohle, D., Greskovits, B., 2019. Polanyian perspectives on capitalisms after socialism, in: Atzmüller, R., Aulenbacher, B., Brand, U., Décieux, F., Fischer, K., Sauer, B. (Eds.), *Capitalism in Transformation*. Edward Elgar Publishing, Cheltenham, pp. 92–104.

Bugarič, B., 2019. Central Europe's descent into autocracy: A constitutional analysis of authoritarian populism. *International Journal of Constitutional Law*, 17, 597–616. https://doi.org/10.1093/icon/moz032.

Chelcea, L., Druţă, O., 2016. Zombie socialism and the rise of neoliberalism in post-socialist Central and Eastern Europe. *Eurasian Geography and Economics*, 57, 521–44. https://doi.org/10.1080/15387216.2016.1266273.

Dorondel, S., 2016. *Disrupted Landscapes: State, Peasants and the Politics of Land in Postsocialist Romania*. Berghahn Books, New York and Oxford.

Gonda, N., 2019. Land grabbing and the making of an authoritarian populist regime in Hungary. *The Journal of Peasant Studies*, 46, 606–25. https://doi.org/10.1080/03066150.2019.1584190.

Kovács, E. K., 2015. Surveillance and state-making through EU agricultural policy in Hungary. *Geoforum*, 64, 168–81. https://doi.org/10.1016/j.geoforum.2015.06.020.

Krastev, I., 2018. Eastern Europe's Illiberal Revolution: The Long Road to Democratic Decline Is Democracy Dying. *Foreign Affairs*, 97, 49–59.

Mihók, B., Biró, M., Molnár, Z., Kovács, E., Bölöni, J., Erős, T., Standovár, T., Török, P., Csorba, G., Margóczi, K., Báldi, A., 2017. Biodiversity on the waves of history: Conservation in a changing social and institutional environment in Hungary, a post-soviet EU member state. *Biological Conservation*, 211, 67–75. https://doi.org/10.1016/j.biocon.2017.05.005.

Nadkarni, M., 2020. *Remains of Socialism: Memory and the Futures of the Past in Postsocialist Hungary*. Cornell University Press, Ithaca, NY.

Nagar, R., 2019. Hungry Translations: The World Through Radical Vulnerability. *Antipode*, 51, 3–24. https://doi.org/10.1111/anti.12399.

Schmidt, O., Theesfeld, I., 2012. Elite capture in local fishery management — experiences from post-socialist Albania. *International Journal of Agricultural Resources, Governance and Ecology*, 9, 103–20. https://doi.org/10.1504/IJARGE.2012.050325.

Scoones, I., Edelman, M., Jr, S. M. B., Hall, R., Wolford, W., White, B., 2018. Emancipatory rural politics: confronting authoritarian populism. *The Journal of Peasant Studies*, 45, 1–20. https://doi.org/10.1080/03066150.2017.1339693.

Stahl, J., 2010. The Rents of Illegal Logging: The Mechanisms behind the Rush on Forest Resources in Southeast Albania. *Conservation & Society*, 8, 140–50.

Szulecka, J., Szulecki, K., 2019. Between domestic politics and ecological crises: (De)legitimization of Polish environmentalism. *Environmental Politics*, 0, 1–30. https://doi.org/10.1080/09644016.2019.1674541.

Thelen, T., Dorondel, S., Szöke, A., Vetters, L., 2011. 'The sleep has been rubbed from their eyes': social citizenship and the reproduction of local hierarchies in rural Hungary and Romania. *Citizenship Studies*, 15, 513–27. https://doi.org/10.1080/13621025.2011.564834.

Tlostanova, M., 2019. The postcolonial condition, the decolonial option and the post-socialist intervention, in: Albrecht, M. (Ed.), *Postcolonialism Cross-Examined: Multidirectional Perspectives on Imperial and Colonial Pasts and the Neocolonial Present*. Routledge, New York, pp. 165–78. https://doi.org/10.4324/9780367222543.

Vasile, M., 2019. Fiefdom forests: Authoritarianism, labor vulnerability and the limits of resistance in the Carpathian Mountains. *Geoforum*, 106, 155–66. https://doi.org/10.1016/j.geoforum.2019.08.001.

Watts, M., 1994. Development II: the privatization of everything? *Progress in Human Geography*, 18, 371–84. https://doi.org/10.1177/030913259401800307.

List of Figures

Chapter 2

Chapter 3

Chapter 4

Chapter 6

Chapter 8

Chapter 9

Chapter 11

Chapter 12

Index

About the Team

Alessandra Tosi was the managing editor for this book.

Melissa Purkiss performed the copy-editing and proofreading.

Anna Gatti designed the cover. The cover was produced in InDesign using the Fontin font.

Luca Baffa typeset the book in InDesign and produced the paperback and hardback editions. The text font is Tex Gyre Pagella; the heading font is Californian FB. Luca produced the EPUB, MOBI, PDF, HTML, and XML editions — the conversion is performed with open source software freely available on our GitHub page (https://github.com/OpenBookPublishers).

This book need not end here...

Share

All our books — including the one you have just read — are free to access online so that students, researchers and members of the public who can't afford a printed edition will have access to the same ideas. This title will be accessed online by hundreds of readers each month across the globe: why not share the link so that someone you know is one of them?

This book and additional content is available at:

https://doi.org/10.11647/OBP.0244

Customise

Personalise your copy of this book or design new books using OBP and third-party material. Take chapters or whole books from our published list and make a special edition, a new anthology or an illuminating coursepack. Each customised edition will be produced as a paperback and a downloadable PDF.

Find out more at:

https://www.openbookpublishers.com/section/59/1

You may also be interested in:

www.ingramcontent.com/pod-product-compliance
Ingram Content Group UK Ltd.
Pitfield, Milton Keynes, MK11 3LW, UK
UKHW020423061026
15625UKWH00006B/379

9 781800 641334